ENDURING SEEDS

Native American Agriculture and Wild Plant Conservation

The University of Arizona Press
Tucson

GARY PAUL NABHAN

ENDURING
SEEDS

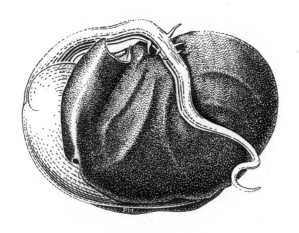

The University of Arizona Press
© 1989 by Gary Paul Nabhan
This edition is published by arrangement with North Point Press, a division of Farrar, Straus and Giroux, LLC.
First University of Arizona Press paperbound edition 2002

www.uapress.arizona.edu

Library of Congress Cataloging-in-Publication Data
Nabhan, Gary Paul.
Enduring seeds : Native American agriculture and wild plant conservation/Gary Paul Nabhan.
p. cm.
Originally published: New York : North Point Press, 1989.
Includes bibliographical references (p.).
ISBN 0-8165-2259-6 (pbk. : alk. paper)
1. Indians of North America—Agriculture. 2. Seeds—North America. 3. Plant Conservation—North America. I. Title.
E98.A3 N33 2002
630'.89'97—dc21   2002018099

Lines from "The Ballad of Ira Hayes" by Peter LaFarge © 1962 are used by permission of the Edward B. Marks Music Company.

An excerpt from *Leaving the Land* © 1995 by Douglas Unger is used by permission of the University of Nebraska Press.

Drawings by Paul Mirocha.

Manufactured in the United States of America on acid-free, archival-quality paper.

13   12   11   8   7   6   5   4

*This book is dedicated to Native Seeds/SEARCH,
its founders, staff, members, and the Native
American farmers who have kept the seeds alive
for centuries.*

CONTENTS

FOREWORD *to Second Edition*

Miguel A. Altieri

The people of the land, the Indian farmers of North America—like their counterparts in Mesoamerica, the Andean region, and the Amazon—have continuously cultivated maize, beans, squash, and other crops for more than five thousand years. One of the salient features of their traditional farming systems is the high degree of biodiversity. These traditional farming systems have emerged over centuries of cultural and biological evolution, and they represent the accumulated experience of indigenous farmers interacting with the environment without access to external inputs, capital, or scientific knowledge. Using inventive self-reliance, experiential knowledge, and locally available resources, traditional farmers have often developed farming systems with yields that have stood the test of time.

In Latin America alone, more than 2.5 million hectares under traditional agriculture in the form of raised fields, polycultures, agroforestry systems, and the like document indigenous farmers' successful adaptation to difficult environments. Many of these traditional agroecosystems constitute major repositories of both crop and wild-plant germplasm. From an agroecological perspective, these agroecosystems can be seen as a continuum of integrated farming units and natural or seminatural ecosystems in which plant gathering and crop production are actively pursued. Plant resources are directly dependent upon management by human groups; thus they have evolved in part under the influence of farming practices shaped by particular cultures and the forms of sophisticated knowledge they represent, as so well described, in the case of North American Indians, by Gary Nabhan in *Enduring Seeds*.

Agronomists, other scientists, and consultants have struggled to understand the complexities of local farming methods and their underly-

ing assumptions. Sometimes, unfortunately, they have ignored traditional farmers' rationales and imposed conditions and technologies that
have disrupted the integrity of native agriculture. After visiting Mexico
at the invitation of the Rockefeller Foundation in the wake of the Green
Revolution, Berkeley geographer Carl Sauer commented:

> A good aggressive bunch of American agronomists and plant breed-
> ers could ruin native resources for good and all by pushing their Ameri-
> can commercial stocks.... And Mexican agriculture cannot be pointed
> toward standardization on a few commercial types without upset-
> ting native economy and culture hopelessly. The example of Iowa is
> about the most dangerous of all for Mexico. Unless the Americans
> understand that, they'd better keep out of this country entirely. This
> must be approached from an appreciation of native economies as
> being basically sound.

Perhaps the greatest challenge to understanding how traditional
farmers maintain, preserve, and manage biodiversity is recognizing the
complexity of their production systems. Today it is widely accepted that
indigenous knowledge is a powerful resource in its own right and is
complementary to knowledge available from Western scientific sources.
Therefore, as suggested by Nabhan, in studying such systems it is not
possible to separate the study of agricultural biodiversity from the study
of the culture that nurtures it. As traditional agriculture disappears in
the face of major social, political, and economic changes, this book makes
an important case: the preservation of traditional agroecosystems must
occur in conjunction with the maintenance of the culture of the local
people. The conservation and management of agrobiodiversity is not
possible without the preservation of cultural diversity.

There is strength in the diversity of indigenous farming systems,
and a major message put forward in this book is that this diversity must
endure for the benefit of the people, especially the poor, throughout the
world. Traditional agriculture is the cradle of agrobiodiversity, plays a
key role in ensuring food security, preserves soil and water, and is resil-
ient to natural disasters, especially climate changes, over which indigenous and peasant farmers have no control.

Ever since the industrialized countries became aware of the ecological services performed by biodiversity, most of which is found in developing countries, the Third World has witnessed a "gene rush" as multinational corporations have aggressively scoured forests, crop fields, and coasts in search of genetic gold. Protected by the World Trade Organization, multinational corporations have freely practiced "biopiracy." The Rural Advancement Foundation estimates that this costs developing countries US $5.4 billion per year in lost royalties from food and drug companies that use germplasm from indigenous farmers' crops and medicinal plants.

Clearly, the world's plants are viewed as raw materials for multinational corporations, which have made billions of dollars on seeds developed in U.S. labs from germplasm that indigenous farmers have carefully bred over generations. So far, biotechnology companies have offered no provisions to pay such farmers for the seeds they have taken and used.

The agricultural systems being developed with biotech crops favor monocultures characterized by dangerously high levels of genetic homogeneity, leading to greater vulnerability of agricultural systems to biotic and abiotic stresses. As the new bioengineered seeds replace the old traditional varieties and their wild relatives, genetic erosion will accelerate. What is ironic is that the push for uniformity will not only destroy the diversity of genetic resources but will also disrupt the biological complexity that underlies the sustainability of traditional farming systems, most of which have provided the raw material for corporate biotechnology.

It is clear that indigenous agricultural strategies favor complexity as a deep ecological rationale. The kinds of agriculture with the best chance to endure are those that deviate least from the diversity of the natural plant communities within which they exist. For this reason, traditional agriculture represents a powerful counterforce against the trend toward a reductionist view of nature and agriculture set in motion by contemporary biotechnology. As emphasized in *Enduring Seeds,* preserving traditional agroecosystems and understanding the ways in which peasants maintain and use biodiversity can speed considerably the emergence and further adoption of sound agroecological principles. Such principles are greatly needed in order to develop more sustainable agroecosystems and biodiversity conservation strategies in both industrial and developing countries. It is the only hope for an enduring agriculture.

Miguel A. Altieri

FOREWORD *to First Edition*

Wendell Berry

Gary Nabhan's work reminds us of what I can describe only as a sort of historical wonder: that we have paid an immense amount of attention to American Indian crops, or at least to some of them, but almost no attention at all to American Indian farming. Books and movies, radio and television have given us images in abundance of the Indian fighting and hunting and participating in various ceremonies, but few indeed of the Indian farming or gardening.

That we should value Indian crops but not Indian farming is probably another instance of our disposition as a people to value substance above form—or, in fact, to destroy form for the sake of substance. We are now destroying our own farms for the sake of our crops, just as we are now destroying our forests for the sake of their timber. As Gary Nabhan puts it, speaking of the Southwestern deserts: "Modern agriculture has let temporary cheap petrochemicals and water substitute for the natural intelligence—the stored genetic and ecological information—in self-adjusting biological communities." And, of course, we have substituted these things also for the human intelligence stored in human communities. "Dispensing with the formalities" is one of the things that, as "forthright Americans," we are proud of being able to do.

This book begins a correction that is necessary and welcome. It is, among other things, a study of the form of American Indian agriculture, the chief principles of which are as follows:

1. Local cultural stability.
2. Local adaptation of cultures and crops.
3. Connections and exchanges between cultures.
4. Interpenetrations and exchanges between the local domestic economy

and its surrounding wild ecosystem, which are necessary to the economy and often useful to the ecosystem.

5. Propriety of scale.

We have known before that these principles were necessary to a sane, continuously productive way of farming. Dr. Nabhan greatly strengthens the argument in their favor by showing them in action in his examples. He thus brings us closer to the day when we will have an agriculture, and thus a culture, truly native to our land. And I like his statement of this goal: "Agriculture is native not merely when native peoples are the farmers. It is fully native when a diversity of locally-adapted organisms function within its fields, lending them yield stability and ecological resilience."

Wendell Berry

ACKNOWLEDGMENTS

The seed for this book germinated a decade ago during a brief discussion with the late Dr. Edward Spicer, a cultural anthropologist who studied and lived among "persistent peoples." Dr. Spicer challenged me to ask, "How have certain cultures managed to persevere in their locally-adapted farming traditions when so many others have abandoned their traditional seeds or have abandoned traditional agriculture altogether?" Soon after that, Wendell Berry and Jack Shoemaker encouraged me to write an overview of native farming traditions still remaining on the North American continent. Although the geographic range covered by this book extends well beyond the Sonoran bioregion with which I am most familiar, these essays are less an overview than an attempt to understand unifying themes in American agricultural history and plant conservation. Nevertheless, the *Enduring Seeds* essays were built on the foundation of scholarship established by Edward Spicer in his 1971 essay on persistent cultural systems, which was extended in his 1980 book on the Yaqui, and by Wendell Berry in his 1981 classic, *The Gift of Good Land*.

Much of the fieldwork and background research for this work has been supported by foundations, including the C. S. Fund, the Agnese Lindley Foundation, the Ruth Mott Fund, Pioneer Hi-Bred's fund for support of public education and research, the National Science Foundation, Institute of Museum Services, and the Tides Foundation. I am indebted to the donors, board members, and the staff of these foundations for their personal interest in this work. My colleagues at the Desert Botanical Garden and Native Seeds/SEARCH have also been wonderfully supportive of this project.

Some of the *Enduring Seeds* essays were originally prepared as lectures to be presented at The Land Institute, the Lindisfarne Association, The Arizona Nature Conservancy, the Center for Plant Conservation, the Tucson

Botanical Garden, and the California Academy of Sciences. If those who heard these lectures barely recognize the contents of this book, no wonder! Some have been rewritten a half-dozen times since then. Others were originally written for *CoEvolution Quarterly, Earth First!, Annals of Earth Stewards, Garden, The Seedhead News, Agriculture and Human Values, American Land Forum, Environment Southwest, The Sunflower, Southwestern Naturalist,* and *The (New) Ecologist.* I thank their editors for printing earlier versions of segments of these essays, and for permission to include them here.

I am grateful to several people for reading the entire prepublication manuscript, and for offering suggestions to improve its cohesiveness and clarity: Wendell Berry, Rod Clark, Kevin Dahl, Greg McNamee, Sheila Helgath, Mark Slater, Kate Moses, Kent Whealy, Jane Cole, and Paul Mirocha. After the manuscript was completed, Jane Cole kindly assisted with the bibliographic references and index, Emily Heckman kept it moving, and Paul Mirocha offered his graphics to supplement the text. Many other individuals read or commented on parts of the manuscript. Some had earlier guided me in the field, or had shared their own farming traditions or primary research with me. I am particularly indebted to these individuals, for they inhabit the landscapes sketched in this book: Gordon Bender, Jim Meeker, Greg LeGault, Thomas Vennum, Karen Reichhardt, Joe Callizo, Richard Pentewa, John Chandler, Subodh Jain, Rick Kesseli, Charlie Rogers, Gerald Seiler, Loren Reisberg, Jono Miller, Julie Morris, Susan Wallace, David Martin, Tom Andres, Emanuel Brietburg, Amadeo Rea, Charmion McKusick, Cora Baker, Mary Baker, Vera Bracklin, George Will, John Doebley, the Lone Fight family, Eric Roos, Suzanne Nelson, Kay Randall, Rob Robichaux, Mark Slater, Rita Buchanan, Lawrence Kaplan, Wes Jackson, Paul Martin, Josh Kohn, Howard Lyon, James Nations, Janice Alcorn, Miguel Altieri, Mahina Drees, David Ehrenfeld, Barney Burns, Laura Merrick, Garrison Wilkes, David Doyel, Charles Miksicek, Donald Graybill, Robert Gasser, Robert Bye, Tim Dunnigan, Linda MacMahan, Ann Zwinger, Wendy Hodgson, Daniel Janzen, David Riskind, Jose Muruaga-Martinez, and Richard Felger. Personal communications from the preceding people will not be intermixed with the following chapter-by-chapter bibliographic essays, but their direct comments to me have been as important as the published literature in writing this book.

For each chapter, I first discuss the more general, non-technical literature which may be of interest to the lay reader. I then cite references which have been quoted or paraphrased, and briefly note other key technical articles which have been interpreted in the text. Many of the particular plant genetic resources mentioned in the text are propagated, maintained, or offered to the gardening public by Native Seeds/ SEARCH, Tucson, Arizona.

PROLOGUE

Enduring Seeds—
The Sacred Lotus and the Common Bean

I.

Cupping my two hands together, I can hold a single seed of each plant that grows on the acres of desert lands where I live. Filling a bottlegourd canteen full with one seed from each genetically-distinct cultivated variety native to this continent, I could also carry around samples of all of the domesticated crop strains known to have existed prehistorically in North America.

When pooled together, these microcosms of life called germ plasm contain more information than is contained in the Library of Congress. In a handful of wild seeds taken from any one natural community, there is hidden the distillation of millions of years of coevolution of plants and animals, of their coming together, coexisting, partitioning various resources, competing or becoming dependent upon one another. In a gourdful of crop seeds taken from fields of Native American farmers, we have the living reverberations of how past cultures selected plant characters that reflected their human sense of taste, color, proportion, and fitness in a particular environment. We also have the germ that generates many stories, many ceremonies, and many blessings.

Seeds. One might simply define them as fertilized, matured ovules, the results of sexual reproduction in plants. Usually, we think of seeds as the enclosed germ of flowering plants, but "naked" ovules such as those of conifers and cycads qualify as well. We tend to focus on the seeds of flowering plants—the angiosperms—because they make up more than half the food that humankind ingests every day. They are the pits of fruits, the grains, the nuts, and the beans that have sustained us ever since we emerged as a species 200,000–500,000 years ago. Today, the fate of these angiosperms is in our hands more than ever before.

II.

In one hand, you hold the seedlike fruits of the sacred lotus, in the other, a common bean. Both of these seeds are elliptic in shape and enclosed in hard, glossy coats. Each is weighty enough for its presence to be felt in the palm of your hand, much the same as a small coin or other historic form of monetary currency—certain seashells or precious stones. Yet in contrast to these other forms of currency, the sacred lotus and the common bean are living, and capable of regeneration. The lotus seed, in fact, can outlive you. And if scientists' dreams come true, beans dunked in liquid nitrogen deep freezes may soon outstrip the human lifespan.

Humankind has long associated itself with both of these seeds, but has affected them in different ways. The aquatic lotus retains its wild-type survival mechanisms; the garden bean does not.

The sacred lotus, *Nelumbo nucifera*, falls into the most primitive flowering plant group still surviving, the ranalean complex, which also includes water lilies, magnolias, and tulip trees. It and its New World counterpart, the water chinquapin, *Nelumbo lutea*, are both aquatic plants. They produce sweetly scented, solitary blossoms that project above their umbrellalike leaves, escaping the muddy waters.

Their seed-bearing receptacle is like a pepper shaker, pocked with openings through which ten to thirty marble-sized seeds protrude. If gentle waves vibrate the lotus stalk, its seeds rattle out. If harder winds break the shaker free, it floats upside down, dropping seeds as it drifts along.

The corky, root-like rhizomes of both lotus species are anchored in the muck beneath shallow lakes and ponds. In North America, native gatherers historically harvested and ate both the acrid "roots" and the seeds of *Nelumbo lutea*. Even today in the Orient, the sacred lotus rhizomes demand high prices thanks to their highly digestible starch, artichoke-like flavor, and medicinal use as a tonic.

However, the sacred lotus did not originate in the Far East. It appears to have been taken there soon after Buddhism was introduced into China from India, where the lotus had already been a powerful religious symbol for centuries: the unfolding flower of consciousness. The Hindu creator, Brahma, is often portrayed as emerging from a lotus flower as it floated on the floodwaters that destroyed an earlier world. Brahma then recreated the universe from the lotus, turning its petals into the hills, swales, and valleys inhabited today. Riding the tide of such powerful imagery, the lotus was dispersed by seed from India and Sri Lanka to China, and then to Japan and Korea.

The lotus arrived in Egypt a few centuries before the time of Jesus of Nazareth, and soon entered Occidental literature and art. It was dispersed to Hawaii and the eastern United States more than eighteen centuries later. The Old World lotus now abounds in ponds maintained by the National Park Service—keepers of America's natural heritage—in Washington, D.C.

One of the pond populations in Washington's Kenilworth Botanical Gardens attests to the fact that the lotus has not lost its wild resilience. It was initiated from a single seed germinated in 1951 by Horace Wester—one of a cache of lotus seeds that scientists currently believe to have been deposited on a now-dry lake bottom in Manchuria over 460 years ago. From this naturally-deposited cache, many seeds have been germinated over the last seventy years. One of them is considered the oldest viable seed in the world.

III.

A Japanese scientist named Ichiro Ohga first drew world attention to a layer of these seeds found at Pulantien in southeastern Manchuria, for he suspected that they were of incomparable antiquity. They were found in an oxygen-poor peat deposit that indicated the presence of a lake where written and oral history had recorded no such water body for more than a century-and-a-half.

The seeds were worn and discolored when compared to freshly-harvested ones. They were lighter, and shrunken. Still, they were solid enough for Ohga to think of attempting to germinate them. A few days after digging them up, he placed them in water, where he left them for eight months without noticing any visible change. Finally, Ohga filed through the hard seed-coat—actually the shell-like skin of the lotus "fruit"—and within four days of additional soaking, all of the ancient embryos sprouted.

Concerned that other scientists would not believe him, Ohga distributed several of the lotus seeds found in the same deposit, and encouraged independent confirmation of their germinability by well-known physiologists. While Ohga remained alive, his colleagues failed to agree on just how old these lotus seeds might be. Three decades passed between the time of Ohga's 1923 excavation and the first direct dating of the seeds by Nobel Prize winner Willard Libby. Libby assumed that he had refined radiocarbon dating techniques well enough to estimate the age of the seeds. His results suggested that they were 1040 years old, plus or minus 210 years. Later, however, a team of skeptical geochronologists disputed this projection. Their analysis returned a date of one hundred years, plus or minus sixty.

More recently, an age estimate of 440 to 460 years has been derived from both geological studies of the lake bed as well as refined methods used to carbon date other seeds from Ohga's cache. In 1942, an Englishman named Ramsbottom successfully germinated *Nelumbo* seeds taken off a dried herbarium specimen in the British Museum, known to have been collected 237 years earlier. Additional museum specimens of lotus have usually germinated, but only after the nearly impervious seedcoat is sawn through, or softened by sulfuric acid.

How does the lotus ever germinate in the wild if it is that difficult to open in the laboratory? Unless abraded by scouring or scarified by acidic organics slowly accumulating on a lake bed, the tough shell of the lotus seed will remain fairly impermeable. Inside, the oil-rich seed maintains a quiescent metabolism over decades, even centuries. Perhaps lotus populations are propagated largely by rhizomes, and new germination occurs only after catastrophic floods of a magnitude that sends even Brahma hiding.

Is it not fitting that the lotus has become associated with immortality? Millions of charcoal-colored lotus seeds now sit in deposits of peat, still respiring, even after their plant populations wane and ponds fill in with sediment. All the while, their images are passed down to us on the sides of prehistoric pottery, in jewelry, in tapestries, and as statuary. The seeds dropping to the bottom of the ponds at Kenilworth Gardens today will probably outlive you and me. Will they outlive our culture, or even our species?

IV.

Now, the common bean is in your other hand. Within the Americas, bean images were part of prehistoric art. In ancient Peru, bean figures were painted with legs upon them. The Inca are believed to have used beans as a kind of communications medium, with runners carrying them long distances down sophisticated roadway systems, transmitting messages from one leader to another. The Inca apparently selected beans of different colors, shapes, and sizes, and gave each type a meaning. A handful of beans transported hundreds of miles from runner to runner could be deciphered by an experienced Inca cryptographer and proclaimed as the "world news" as it was known at that time.

Even though the bean family as a whole is known for its relatively long seed viability, commonly cultivated garden bean seeds die young. A batch of domesticated Navy bean seed cannot normally germinate if it is stored a dozen or more years at room temperature. Over its eight to ten thousand years of being genetically domesticated and cultivated by American Indians, the bean has lost much of its hard-seeded qualities when compared to its wild ancestors growing in Latin America. The seedcoat has become thinner, more easy to break or deteriorate as moisture is imbibed, even as the food reserves within it have increased in volume.

Archaeobotanist Lawrence Kaplan has devoted thirty years to contemplating bean domestication in the New World. Collaborating with archaeologists all over the Americas, Dr. Kaplan has received bean seeds and empty pods excavated from prehistoric dwellings and food caches. He is routinely asked to identify them, measure them, and determine whether they appear to be wild or domesticated. After having amassed hundreds of records of common beans, Kaplan noticed a startling trend. The pattern of American bean domestication conflicts with that found in the Old World. In the Mediterranean region, gradual seed size increase through time occurred as pulses such as peas, lentils, garbanzos, and favas were progressively domesticated. In the Americas, beans appear to have been transformed more quickly than the eye of an archaeologist can see.

Although small wild beans with explosive pods are found in a few Mesoamerican sites, the first clearly domesticated beans in the archaeological record are nearly the size of modern-day bean crop varieties. The transition from small to large seeds that accompanied bean domestication is simply not represented in the American archaeological record. This change, Kaplan argues, must have come early and rapidly in the history of New World agriculture—prior to 8,000–10,000 years ago in South America, and as early as

ᐟ the Mexican highlands of North America. And Kaplan
ᴄh a rapid evolutionary change may have taken place.
ᴜs have seedcoats almost as tough as that of the lotus. Some have
ᴄs that harbor chemicals to discourage consumption by tiny bruchid
beetles. These beetles feed on legumes throughout the Americas, seeking
out larger seeds and laying their eggs on the surface of the mature ones.
When their eggs hatch, the larvae penetrate the seed. They consume the
carbohydrate- and protein-rich food reserves within the seed, and often rav-
age the embryo, destroying its capacity to germinate.

Other factors being equal, the smaller the seed, the less likely it is for the
beetles to select it for egg deposition. Recently, researchers have found lectins
and pigments in wild bean seedcoats that function as feeding deterrants for
beetles, yet Kaplan believes that small-seededness alone may offer enough
protection. In any case, the tiny-seeded ancestors of common beans had a
working set of protective mechanisms.

Rather suddenly, when early bean gatherers began to remove beans from
the environments where beetles proliferated, the pressures to remain small
and tough were released. By bringing gathered seed home to dry by mild
heating for storage over the winter, the beans were guarded from the effects
of bean beetles. Kaplan suggests that rapid "selection for large seed size
could have taken place . . . as a result of simple storage practices by non-
agricultural, gathering peoples." Subsequently, with sowing, other colors of
beans were selected as "markers," and some did not have or need the pigments that served as feeding deterrents for so long. Over a rather short period of time, beans were allowed to diversify into new shapes, sizes, and
colors that had been unknown in the wild during the previous thousands of
years.

There was only one hitch. These newly domesticated beans could not
thereafter survive on their own in the wild. With a relatively thinner seed-
coat, the cultivated beans could more readily take up moisture during hot,
humid months, and were easily spoiled. At room temperature in a semiarid
or subhumid climate, fifty out of every hundred beans will lose their viability
within six years' time. For a bean variety to survive, its seeds must be planted
every few years. Its destiny has become inextricably linked with the human
cultures that draw upon it as a favored food.

That is the irony—or better, fallacy—of the widespread folklore regarding "prehistoric Indian beans" said to have come from archaeological sites in the American West. Amateurs claim that they were given such beans by friends, who obtained them from other friends, who found them within pots left in caves or cliff shelters. Upon planting them, the ancient seeds supposedly germinated! Hundreds of tall tales of this sort have been heard in the Southwest, sometimes about Jacob's Cattle beans, in other instances about Aztec, Anasazi, Moctezuma, or Gila Cliff beans. Virtually none of the stories record the exact place, sedimentary stratum, or cultural provenience from which the beans were derived.

To my knowledge, none of the "original" germinable caches have been directly radiocarbon-dated. Most often, the beans actually came from contemporary gardens situated near prehistoric ruins, or from historic monuments where modern analogs of pre-Columbian crops are grown for educational purposes. In most cases, well-meaning individuals simply pass on a few seeds and a vague story to other curious folks. However, a few snake-oil peddlers have gotten involved. In Silver City, New Mexico, such beans are portrayed as mementos of the Old Southwest, and sold for one dollar a bean or one hundred dollars a pound. It seems that people *want to believe* that the very crops so dependent upon man for survival from year to year can somehow last millennia without human intervention.

The only way that the diversity of cultivated beans will persist on this earth is if human cultures care wisely for them. Where Indian farming has persisted in North America, families frequently grow four or five different kinds of beans every year. Even today, it is possible to find villages that harbor ten to eighteen "land races" or locally-adapted bean variants, most of which have been passed generation to generation for no less than a century.

Still, such bean-growing areas seldom surface in statistical reviews of legume production in North America. The tonnages of beans that reach the supermarkets across the continent are from a very few highly uniform microregions, where one, at most three, bean cultivars dominate the fields. Under the present circumstances, the U.S. National Research Council has predicted that an epidemic could easily devastate the entire Western dry bean industry. Diseases or pestilence could also have catastrophic consequences if they ever hit the Eastern green bean production areas at the wrong time, for they too are supported by a stringbean-thin genetic base.

The thinning-out of bean diversity has happened in a matter of a few generations, at least in places such as New York state. As late as 1908, the Iroquois nations of upstate New York were growing no less than sixty varieties of beans, including Cornstalk, Wild Goose, Marrowfat, Hummingbird, Wampum, plus Kidneys and Cranberries of several colors. Their seasonal festivals included a Green Bean Ceremony; with maize and squashes, beans were one of the "Three Sisters," also called *diohe'ko*, meaning "these which sustain us." One Iroquois legend, "The Weeping of Corn, and Bean, and Squash People," tells of an elderly woman who hears the crying of crop plants that had not received proper care. As she begins to weep as well, other villagers come to the fields. When they hear the cause of her sorrow, they join in the wailing. Stephen Lewandowski interprets this story as reminding people today that crop failures of the Three Sisters were caused by the community neglecting their duties.

Today, the Green Bean Ceremony is but a memory, and most of the sixty Iroquois beans are gone. Gone too are most of the 260 other kinds of common beans found on the thousands of small farms which once dotted New York. They have been replaced by just two introduced varieties on over three quarters of the state's dry bean acreage, and by a handful or so of streamlined, college-bred green beans in the gardens and truck farms elsewhere in the state.

Once the last farmer decides not to plant or harvest a variety again, the remaining seed might be harvested and eaten until none remain. If there are plants drying in the field when the market changes, they may simply be plowed under, seeds and all. The beans left in the soil cannot persist the way the lotus can that is buried in organic muck, or wedged into a peat layer. The bean will not have the wild-type resilience or a low-oxygen environment in which to take refuge.

The only factors that have stood between some old-fashioned beans and their extinction are the kindness and curiosity found among exceptional individuals, or encoded within the values of entire cultural communities. How else do you explain the 900 or so heirloom bean samples that John Withee grew out "as a hobby" in New England over the past few decades? Is there any other way to account for "the Bean King of Michigan," Ralph Stevenson, who told a journalist that he planted two hundred and fifty kinds of beans in his garden each year because "I just like to see them grow"? Except for his plots near Tekonsha, an old bean-producing center in southern Michigan, "there isn't hardly a bean raised around here" anymore.

The same tenacity was found in the late Burt Berrier's thirty-year quest for beans across the West, conducted while he worked his way through small towns as a John Deere farm machinery demonstrator. He collected the beans that Brigham Young reputedly carried with him from the Midwest to Utah. Berrier also guarded a native Southwestern bean given to him by a Navajo woman. She had been able to grow it only by hauling pots of water for irrigation across miles of desert lands.

V.

The lotus and the common bean. The lotus might persist regardless of whether humans were ever again to paddle into an Oriental lake. Common beans, on the other hand, are entirely dependent upon the planters, threshers, and storage shelves of humankind. Our destinies are intimately intertwined. If beans persist another four hundred and sixty years, it will surely be because of the curiosity and care—and perhaps the sense of community—of folks like you and me.

Part One

A New World Perspective

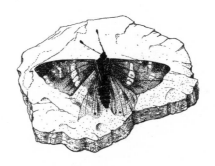

CHAPTER ONE

The Flowering of Diversity

I.

I scramble through a Rocky Mountain meadow, heaped with piles of snow, searching for stalks of flowers, clusters of seeds. Here and there, where an earlier winter blizzard had not beaten them down, spent seedheads still stand above powdery drifts. Breaking through the icy surface, a few have enough features remaining for me to identify them: smooth goldenrod; cow parsnip; Colorado thistle. Only one plant I encounter hangs on to last fall's seeds, glistening black like coals in a bed of whiteness. The lone seed-keeper is a wild onion. In contrast, most plant matter is already plastered against the ground, down below this crystalline cover.

Closing my eyes to escape the blinding brightness of the present, I try to imagine what the meadow must have been like the summer before. Some four hundred plant species blessed this valley then, and fifty were wildflowers that bloomed profusely. They lured a dazzle of bumblebees and butterflies to their nectars.

Today, insects are nowhere to be seen. Where I stand, wildflowers have been frozen back, snapped, or beaten down by the wind and snow. It dawns on me, then, that below my feet, another record of past flowerings lies frozen. Frozen in time.

I should tell you that this Colorado valley is named *Florissant*. It's a place that gives fresh meaning to the term "flower beds." Located forty miles northwest of Pike's Peak, this spot was christened *florissant*, or "blooming," by an early settler. They say that he was awed by the burst of wildflowers that he saw there in 1870. A year later, a young scientist named Mead came to the Rockies collecting butterflies. He heard that great petrified tree stumps had surfaced nearby. He rode to Florissant and turned up remains of redwoods turned to stone. Mead found fossils of insects and flowers as well.

It is not hard to see why paleontologists feel that Florissant is the Rosetta Stone of life in North America as it was 34 to 36 million years ago. Within the tissue-thin layers of paper shales, impressions of a hundred or so species of plants have been found in near-perfect condition. Fig leaves, maple samaras, hickory nuts, redwood sprigs, mesquite pods, pine cones, and magnolia flowers are among the many lifeforms that have left their mark. For the various fossil flowers, delicate but replete with pollen, anthers and petals intact, they lack little more than colors and odors.

Attractive hues and fragrances must have been present in full force at one time, for many flower-loving animals are found in the Florissant beds. For every flowering plant species identified so far, ten kinds of insects have left their impressions.

These fossilized plants and animals reflect a subtropical climate, a different place. Tectonic plates have moved, and it appears that the valley was once situated 5300 feet lower and six hundred miles to the south.

Then, around thirty-five million years ago, volcanic eruptions began that carried mud and lava down the valley, forming a dam and a sickle-shaped lake behind it. A half million years later, more violent eruptions from volcanoes nearby covered the lake with layer after layer of ash, inundating the surrounding mix of tropical- and temperate-derived plants. These plants and the animals dependent upon them became entrapped in the settling cloud, died, and drifted to the lake bottom.

Among the lost ones were dragonflies, bumblebees, hoverflies, butterflies, beetles, moths, and syrphids, all pollinators of plants laid to rest in the sediments. They were pressed within the layers of ash-laden muds that later became the laminae of shales. Their wings, antennae, and mandibles, even the hairs on their legs, were preserved in fine detail.

I reflected on the shapes of nectar-feeding insects in the Florissant fossils, wondering: Is it possible to match prehistoric pollinators with complementary flowers found in the same volcanic bed? All the pieces to this puzzle have not yet surfaced, but it is clear that insects were drawn into particular floral designs by the time of the ash fall thirty-five million years ago.

The evolutionary matching of flowering plants (angiosperms) with pollen-carrying insects occurred fifty to eighty million years before the Florissant strata were set down. Both flying insects and land plants had existed for more than a hundred million years, but their ecological partnership did not begin until the mid-Cretaceous period. Then, as natural historian Philip Regal envisions it, "an ecological revolution 113 to 85 million years ago" led to the displacement of previously dominant tree ferns, horsetails, and cone-bearing gymnosperms, so that "angiosperms now dominate much of the earth's land surface."

To gain a sense of this revolution, I must leave behind the beds at Florissant, and wander further back into time. I must visit other locations where plants and animals have emerged as fossils in the Americas, continents that emerged as distinctive entities themselves only after the Pangean supercontinent broke up about 180 million years ago. No single site captures an entire evolutionary sequence, and few offer snapshots as sharply focused as those at Florissant. Instead, the clues must be gathered from a number of ancient American fossil beds, from Martha's Vineyard, down the coast to the Potomac in Maryland, and inland to the Dakota formation in central Kansas. The revolutionaries have left their tracks in widely-scattered localities. These paleontological nooks and crannies are the true historic sites of the revolution that shaped what our own habitat would become.

Darwin, among other scientists of the last century, pondered the dramatic historic change in the planet's flora that had happened around the Cretaceous period. And yet this biotic revolution was so poorly understood during his lifetime that he called it the "abominable mystery." If today's paleoecologists are correctly interpreting the dynamics of this turnover, the course of life swerved when flowering plants and their insect pollinators, followed by avian and mammalian seed dispersers, all became interdependent.

A Florissant ash fall, a pollen-laden carpenter bee, and the swallowing of a sticky seed mass by a hungry mastodon: these acts seem distant from our daily lives. If, some time ago, seed plants emerged with new modes of pollination and dispersal, what relevance does this have to our own history? The answer is profoundly mundane. Over eighty percent of what we have eaten for thousands of years is angiosperm tissue, in the form of grains, roots, leaves, and fleshy fruits of plant families that did not exist until this revolution changed the world.

In one way, a retrospect on this revolution must humble us. Few of the edible, nutritional characteristics of the seed plants that now sustain us evolved for our benefit, under selective pressure from our forebears or through conscious breeding by scientists. We are literally living off the fruits of other creatures' labors—those of the birds, bugs, and beasts that loosely coevolved with seed plants over the last hundred million years.

II.

I walk through abandoned croplands in southern Arizona, and upon reaching a concrete-lined canal at a field edge, I sit to consider my serendipitous harvest. Cockleburs and stickleaf are plastered to the legs of my jeans. The bony extensions of a devil's claw capsule have wrapped themselves around one of my ankles. Wild barley and three-awn spikelets are stitched into my socks. I have become one more animal used by seeds on their way to new homes.

To learn how seed plants gained the features that are manifested today, we must go back to their "flowering" in geological time. Early in their evolution, seed plants did not attach themselves to animals as readily as they do today. In particular, the large-leaved, herbaceous vegetables and fruits we now eat do not look, smell, or taste much like early angiosperms. Although streamlined, single-headed sunflowers and hybrid maize still share some features with their forerunners, they are a far cry from early angiosperms. Both their life histories and their spatial mixtures with other plants have been radically altered by man, leaving us little notion of the conditions under which flowering plants evolved.

So let us leave the weeds, the big-headed sunflowers, and maize behind in their furrows. Instead, imagine shrubs with cycad-like fronds and magnolia-like flowers. These grew within relatively open stands of conifers, or colonized somewhat dry habitats that had been disturbed by avalanches, lake level changes, or river meanders. Their unspectacular flowers had faint but sweet fragrances—musty, fruity, fermented, perhaps wine-like. These odors, according to Louisiana biologist Leonard Thien, originated from plant chemicals that deterred animals from feeding on leaves, stems, and seedstalks. In flower petals, however, the same chemicals were more dilute than in vegetative growth. Certain insects were stimulated rather than repelled by them.

Once insects began to be attracted to flowers, they found them to be the sources of energy-rich food in the form of pollen, nectar, and ovules. They also took shelter in the flowers, for there it was safe enough to carry on mating. Even today, certain tropical beetles seek safe harbors within the petal-like bracts of *Cyclanthus* inflorescences. If this protective cover is removed during daylight hours, as floral ecologist J. H. Beach did in Central America, the beetles are quickly eaten by lizards and toucans. When left undisturbed within the flowers, the beetles pollinate *Cyclanthus*.

Leonard Thien has asked which lures came to flowers first: those of smell or those of sight. He believes that early flowers must have used aromas to advertise themselves, for it was long before visual cues such as bright blotchy bull's eyes and striped landing pads came into existence. In an ancient plant known as *Zygogynum*, Thien has found that fragrance arises before floral pigmentation appears. Take away the odor-producing glands in certain flowers, and beetles lose interest.

Once insects began to frequent flowers for shelter and food, rapid specialization followed. Floral morphologies developed that allowed the deposition of pollen on insect body parts. Insects that travelled between flowers in the same population then became efficient agents of pollen transfer and floral fertilization. These insect groups shifted from being mere phytophagous (plant-eating) parasites, to symbionts that accrued consistent rewards for their role in pollination.

Flying insects often follow the same trails or "traplines" through the forest, visiting widely separated plants of the same fragrance. Thanks to the skill of these insects, plants that are rather rare in a diverse forest community can still produce outcrossed offspring, even when the parents do not live next to one another. Such mixing of genes between patches of trees may have been a prerequisite to the flowering of diversity.

And yet, cross-pollinated plants could hardly afford to be spaced too distant from others of their kind until widely foraging seed dispersers came into existence. Before there were animals to carry propagules away from the shadow of a mother plant to similar microhabitats elsewhere, pests and predators could easily build up in number. Pests could easily waste any seed crop resulting from these natural "monocultures." Once a variety of birds and mammals began dispersing seeds in different manners, the mixed forest arose. Philip Regal feels that the overlap of wide-ranging seed vectors with pollen-carrying trapline travellers was what really allowed diversity to flower. From then on, outcrossed variants of flowering plants were able to explore new ecological niches.

The relay of pollinators and seed dispersers carried plant evolution toward greater diversity in the early Tertiary Age, about seventy million years ago. By that time, different flowering plants had specialized their nectars and fragrances, and their seeds and fruits as well. In response to the varied fauna of birds, bats, and land mammals attracted to them, some fruits and seeds increased in size. Their larger food stores offered longer survival on forest floors where light penetration was poor. They soon had the edge on conifers for occupying ground where they had formerly been excluded.

Early Tertiary seeds took on a greater spread in seed sizes, if Yale paleobotanist Bruce Tiffney's measurements reflect the true story. Seed plants were suddenly specializing in one habitat or another. Perhaps small-seeded shrubs colonized open patches, while large-seeded trees worked their way up into the forest canopy.

The Rocky Mountain meadow, the ponderosa pine stand, and the dense shrubby edge in between—vegetation takes particular shapes on different landforms at Florissant today. And yet, prior to the revolution, such distinctive communities probably did not exist side by side. Conifers may have dominated all but the most disturbed sites.

Diversification of flowering plant communities gained momentum only when angiosperms could hold their own against needle- and frond-bearing gymnosperms under some conditions. The latter's dependence upon wind pollination became a liability. A wind-borne pollen grain had a tough time finding the cone-bearing tree in a land lush with insects and angiosperms.

By the late Tertiary Age, the vegetation of the earth had shifted from the dominance of trees whose seeds had naked ovaries—the gymnosperms—to the clothed and covered seeds of the plants that form the core of our diet, the angiosperms.

Just as flowers were selected to attract, caress, and cajole insects into carrying pollen, seeds were evolving distinctive signatures that birds or mammals could recognize: clothing, coloration, tastes, and nutritional rewards. These creatures, thus attracted to the seeds, would then disperse them to appropriate microenvironments. When handled, swallowed, and passed through guts, seeds fell under new selection pressures for shape and durability. Some shrubs came to produce fat, calorie-rich fruits to attract animals. The seeds inside needed to be scarified by stomach acids or imbedded in moisture-laden manure to hasten germination. Others manufactured fruits with sticky, sugary pulp that temporarily glued seeds onto fur or feathers. Hooked, horned, winged, or barbed appendages enabled other fruits to travel along on the animals' fetlocks, landing them in distant locations.

Animals unwittingly dispersed the propagules of angiosperms to other patches where they could germinate. The progeny proliferated, and in time diverged from their ancestors. Thus flowering plants radiated into many rare and beautiful forms following long-distance dispersal, sometimes through true coevolution with their allegiant creatures.

III.

Plant diversity had truly flowered. While animals worked on precision pollination and seed dispersal, a number of other reproductive advantages were emerging. The energy-rich triploid endosperm of flowering plant seeds gave them an edge in ripening quicker and in storing sufficient food to maintain the seedling until it could begin to produce its own supplies. Angiosperms also became more capable of faster growth rates and more varied periods of reproduction.

It was not just the additional protection of a pericarp layer surrounding the seed that offered angiosperms an edge over gymnosperms in many environments. New leaf shapes and growth forms, from herbaceous vines to annual grasses and giant palms, had come into existence. These novel architectural forms packed together into plant communities that began to modify the microclimate and to condition soils, and so they made their physical environment more favorable for the growth of future generations.

With such newly-acquired characteristics, flowering plants moved out of the primitive evergreen forest composed of only one canopy stratum, to create the many-layered tropical deciduous forests and the patchwork of the savannas. In mixed forests and savannas, different kinds of plants competed for the same resources, to be sure, but some of them also gained benefits from their neighbors. Taller trees could reduce the extreme temperatures, water loss, and wind damage to which understory plants are susceptible. Plant neighbors frequently share beneficial microorganisms with one another, and thereby gain access to nutrients that might otherwise be unavailable. One plant may provide physical support or a refugium from prey for another. Where a diversity of lifeforms can be found, there may be hidden several subtle but beneficial relationships between plants; these may control the structure of vegetation as much as competition does.

Such lifeform diversity, replete with mutualisms, is particularly evident in subtropical canyons edging North American deserts. There, where unpredictable rainfall keeps any single plant strategy from outcompeting others every year, a dozen forms of plant architecture coexist. Their roots interlock like pieces of a puzzle, partitioning the limited rainfall.

Flowering plants of different shapes and sizes were soon spanning a range of climates, from those receiving two hundred inches of rain per year, to those getting by on a meager two. At one extreme, plants invested their energy in deep roots, massive leaves, and tall trunks reaching toward the light. At the other, in drier environments, some plants developed extensive but shallow roots and small leaves with means of reflecting away light and reducing heat loads. If any single factor forced flowering plants to diversify their survival strategies, it was the variation in climate across the face of the earth. Rainshadow deserts within a few miles of cloud forests drive this point home.

IV.

I once played hopscotch, jumping from one world of plants to another, where the "Gran Desierto of Sonora" enters Arizona along the U.S.–Mexican border. I started off on the edge of a black lava flow and found one set of plants, including squat cacti and resinous shrubs with roots invading every crevice where water might be held a moment. With a skip and a jump, I had dropped down into dunes abutting the *a'a* lava, where the plants endemic to shifting sands caught my eye: desert lilies and other flower-producing bulbs adapted to loose soils. Some of the dune plants—a shrubby Mormon Tea and a gangly milkweed—had rock-loving analogs on the lava above. I then hopped in another direction onto a dry playa, where salt-tolerant shrubs and tenacious bunchgrasses of the talc-like silt flats spoke an altogether different language.

Hold the climate constant, but vary the substrate, and you will still find diversity. The plants that thrive on serpentine soils are seldom the ones which endure gypsum dunes or cover clay shales. Where landscapes of different geological origin become juxtaposed, botanist Arthur Kruckeberg has observed, "the opportunities for events leading to speciation" are manifold.

Over a range of environments, wild seeds sorted out into different shapes, colors, and sizes, each functional in a particular set of conditions. Some came to match the color of the soil they grew upon, so that they would be camouflaged and hidden from animals. Other seeds and fruits took on bright scarlet and orange hues in order to attract birds that could see red, while escaping the notice of color-blind mammals having guts with the capacity to destroy seed viability. Certain grains gathered themselves together on rachises in a way that let some fall free while others were gobbled up by turkeys or other land birds. Fruits the size of softballs were cracked open by mammals the size of Mack trucks, while minuscule marsh plant seeds could be carried for thousand-mile rides on the mud stuck to duck feathers.

Over an eighty- to one-hundred-million-year period, an enormous variety of fruit and seed types evolved, largely under the dispersing power of wind, birds, and mammals. Fruits and seeds vary in shape, size, stickiness, ease of handling, hardness, nutritional content, color, longevity, and taste. It took plant anatomist Crocker nine hundred pages to survey the basic seed types of the dicots, the flowering plant group that begins life with two seedling leaves. Monocot seeds, including those of grasses, lilies, irises, and orchids, would require several hundred pages more.

Of an estimated 300,000 angiosperms that may exist in the world, perhaps 130,000 of these plants reside in the Western Hemisphere today. No one can accurately guess how many of these plants inhabited the Americas when Clovis hunters first encountered this bounty around twelve to thirteen thousand years ago. Even if Clovis hunters were not the first to set foot on the American continents, they are nevertheless considered the first American culture. Some archaeologists argue that other people may have discovered America earlier, for human hunters had been expanding their ranges toward the far reaches of the planet for nearly 100,000 years. Yet if earlier Americans existed, they left no widespread artifact tradition as a legacy to us on this continent. The Clovis tool kit is the first consistently recognizable cultural heritage left in the New World. This culture was derived from that of Beringian folk who flourished within the arctic refugia of eastern Siberia during the last Ice Age. They had become extremely successful large animal hunters by the time they crossed the Bering Straits into North America.

As the Beringians reached the southern tail of an ice-free corridor near the present site of Edmonton, Alberta, they entered a land full of large game that had never known a human predator of their kind before. The opportunity was unique. As prehistorian François Bordes put it, "There can be no repetition of this until man lands on a hospitable planet belonging to another star."

V.

What had been true wilderness became a human-inhabited land. Imagine, as archaeologists Whittington and Dyke have, that the earliest Americans ". . . faced a wide-open, extremely rich continent with an abundant megafauna, so they hunted and reproduced at very high rates." And then, as the archaeologists understate it, "the human population grew and put pressure on resources," forcing the overkill and extinction of many large mammals on a continent-wide basis.

This vision, or nightmare if you will, has been refined into the Pleistocene Overkill hypothesis by Paul Martin. Martin, an eclectic ecologist, has not convinced all paleoecologists of a prehistoric overkill in the Americas, but he has skillfully dealt with a variety of counterarguments during three decades of debate. Martin contends that within a relatively short time—1000–2000 years—the human population growth fueled by this carnage spread to the far tip of South America. Along the way, Clovis hunters and their progeny had driven into extinction 68 genera of animals weighing 100 pounds or more.

There is no convincing evidence that these big game hunters spent much time gathering plants to supplement the meat in their diet. That is ironic, because by then the insects, birds, and mammals that preceded them had created ecosystems with more plant diversity than has existed since. The mammoth-sized megafauna had influenced particularly the evolution of a wide variety of fleshy fruits: drupes, pods, pomes, berries, and pepos. More than fifty genera of Neotropical plants have characteristics that suggest that they loosely coevolved with megafaunal fructivores. Most of these plants have huge fruits with tough, dull-colored skins surrounding a fibrous, carbohydrate-rich pulp and gut-resistant seeds. Curiously, these seeds seldom germinate in the shadow of their mother plant. They must be dispersed into the open, or passed through a mammalian gut, before they sprout.

The rather sudden loss of so many seed and fruit eaters on the continent may have left many American plant species without their loosely coevolved dispersal agents. Woody perennials such as custard apples, calabash trees, cacaos, jocotes, zapotes, prickly pears, large-nutted palms, and perhaps even wild gourds became anachronisms. With few big-mouthed, large-toothed, long-necked animals around to pull their fruit down, there were fewer ways that these fruits could be cracked, swallowed, transported, and excreted. When it came to reaching safe sites for regeneration, these fruits were as stranded as a fish in a tree.

VI.

The human population in North America had grown rapidly, perhaps to a size of one million, by the time of big game extinction. With most huntable herbivorous species gone, the human population on the continent may have had its growth slowed temporarily because of this reduction in the land's carrying capacity. The remaining people scrambled to make use of the anachronistic fruits and seeds left behind by the megafauna, as well as of many other plant species and smaller game.

These nonagricultural peoples may have sought other ways to concentrate food resources with relatively little effort. There is widespread evidence that hunter-gatherers often took advantage of caches of seeds, nuts, and bulbs stored by the animals that they were hunting. Historically, Pawnee and Winnebago women would appropriate pounds of ground beans found in the stores of meadow voles during the winter. Lewis and Clark saw Northwestern Indians robbing muskrat nests for the tubers of aquatic arrowheads, known as tule potatoes. The Seri and Yaqui Indians of the Sonoran Desert would unearth caches of mesquite pods sequestered by pack rats. While gathering up the pods from the rodent's middens, they would catch, kill, and roast as well any pack rats they happened to find.

Nonagricultural peoples have almost always had the knowledge and skills to burn, clear, or flood lands to favor the production of wild food plants. And yet, until their populations burgeoned, most of them did not use these labor-intensive techniques for managing plants to produce more food. After the Clovis hunters, however, some cultures began intentionally sowing and selecting fruits and seeds. This happened 10,000 or so years ago in Mesoamerica, and 6,000–7,000 years ago in North America.

In 1970, geographer John Alford advanced the hypothesis that faunal extinction and the origins of New World agriculture are temporally and causally related. He maintained that abundant game and the need to follow animal herds would have served as a disincentive for settling down to tend the plants on a particular patch of land. Alford observed that the earliest agriculture began south of where the remaining bison species (which had escaped extinction) still roamed. Around 7,000 B.C., bison herds were restricted to the regions north of Mesoamerica, so that some plains and woodland dwellers had ample incentives to continue their big game hunting traditions.

To the south, in the arid subtropics and wet tropics of Mesoamerica, cultures were left without large herds of grazers and browsers, but they did have the gourds, calabashes, yucca fruit, and other plants that evolved under these animals' influence. Did these Neotropical dwellers take up the gourd, and its seeds, to become their agents of dissemination and increase? Did we, as a species, become one more way that seeds reach suitable sites?

During the winter of 1987, I became possessed by these questions. I first listed all of the Neotropical plants presumed to have been dispersed by now-extinct megafauna. I then found that nearly one hundred plants on the list have served as foods for Central American cultures. At least forty-five of these wild plant species have been encouraged, protected, or transplanted by Mayan peoples in Central America. Could it be that Mesoamericans were propagating and promoting "wild" fruit trees in the tropics long before they began field agriculture? Who could have passed up the large sweet fruits of the tropics, especially once the meat supplies had dwindled?

I wondered, then, if many of the fruits that coevolved with American megafauna are listed among the plants now cultivated as crops in the Americas. To my surprise, I found that fifty-five plant species fall into both categories. They were probably found in the oversized dung of gomphotheres, ground sloths, and mammoths, and are now found in tropical orchards or dooryard gardens of the human successors of gomphotheres, of sloths, and of mammoths. Avocados, mamey zapotes, nanches, calabashes, jocotes, and squashes were found in the homes and hearths of humans as far back as the Mayan Pre-Classic period. Human dependence upon them may hypothetically predate the first cultivation of corn and beans. And the time span between the last American appearance of a mammoth and the first human-gathered, oversized fruit is rapidly shrinking.

I now doubt that agriculture could have begun in the Americas before most of the megafauna was gone. It is easy to imagine what havoc a hairy elephant could wreak in your field of squashes. Perhaps it is not so ironic that the second flowering of plant diversity on this continent occurred after the demise of large wildlife diversity.

That second flowering, of course, is of the cornucopia of domesticated plants that the Americas have offered the world. Once the ground sloths, giant bisons, horses, and elephants were extirpated, their coevolved seed plants soon came under the aegis of human craft and curiosity. They then diverged into a myriad of cultivated varieties, land races, or crop ecotypes.

It is during this period—8,000–10,000 years ago—rather than in the age of the "invention" of hybrid corn and tractors that American agricultural history begins. From this time onward, like the bugs, birds, and beasts before us, Americans have had to devote themselves to pollinating and dispersing the plants that nurture and shelter them.

Except for the largest ones, the animals have not left our fields. Their influences are still imbedded in the food we eat. They persist as pollinators, as soil aerators, and in some places, as beasts of burden for our harvests. Their impressions are left in the shapes of the fruits on our kitchen tables as clearly as the floral impressions set in fossil "flower beds" at Florissant.

CHAPTER TWO

Diversity Lost:

The Wet and the

Dry Tropics

I.

The day I first passed through it, I paid little attention to El Rancho, for it seemed to be a desultory pit stop, unrelated to the purpose of my journey. It was no more than a junction where my trajectory would pivot, to head northward into the Mayan highlands of Guatemala. Like many before me, I paused there only a moment amidst the dust, smoke, and motor roar, hardly perceiving El Rancho as a place in and of itself. Later, speaking with friends who had worked for some years in Guatemala, I realized that it is no mere highway intersection. El Rancho is a crossroads between life zones, lifeways, and different paths that Guatemalans may take, that will make or break the health and wealth of their land. It's a juncture, where we see wet and dry worlds joined, and where it's painfully clear that the vegetative fabric of each of these worlds is coming undone.

Located in Guatemala's Oriente region, fifty miles from the capital city and sixty miles from the Honduran boundary, El Rancho lies where two major trade routes converge: Central American Highway 9 and National Highway 17. As I walked the highway margins and peeked into makeshift market stands, I saw workers unloading items that had come in from one direction, as vendors sold these goods to those coming from another.

It was a menagerie of raw materials, ripening fruits, and finished products. Rum, sugar cane, candies, mangos, tamarinds, cashews, fuelwood, bananas, beans, limes, beer, nanches, henequen nets, jocotes, turkeys, Tupperware, and baseball caps were all bought or bartered for. Despite the many products—hand-nurtured or manufactured—El Rancho seemed washed-out compared to the colorful native markets of the highlands or the port towns of the Caribbean coast. Amid all the dust and smoke of the diesel trucks and woodfueled roadside cafes, it was easy to overlook the geographic dynamic that made El Rancho so intrinsically suitable as a crossroads.

Friends had brought me to El Rancho from the southwest, from Guatemala City and the mid-elevation towns dominated by *ladinos*, the people of mixed Spanish and Indian heritage whose lives are being rapidly modernized. Winding through miles of smouldering slash-and-burn fields and pastures, we finally dropped down into the rainshadow area of the Oriente. In contrast to the lush uplands, the low, dusty hills surrounding El Rancho spoke of aridity.

The Oriente was once dominated by dry subtropical and tropical forests, although four hundred square miles of it has been classified as arid subtropical thornscrub. Its vegetation looks much like the dry subtropics elsewhere in Latin America. The highly seasonal rainfall varies greatly from year to year and place to place, ranging from sixteen to forty inches. Much of the year, soil moisture is limited, since annual evaporation exceeds rainfall by ratios ranging from 3 to 1, to 16 to 1, depending on the locality.

Fortunately for farmers, the Rio Montagua drains the uplands of the west to bring the possibility of limited floodplain irrigation to El Rancho and agricultural towns downstream. Dropping more than three thousand feet to El Rancho's six hundred feet, the Montagua then meanders eastward, eventually flushing into the Gulf of Honduras.

To the north and northwest, mountains shrouded in clouds hold the rains that seldom reach the parched slopes skirting El Rancho. There, from 4500 to 9000 feet above sea level, humid rainforest covers the lateritic soils. The vegetation itself is veiled by mists during much of the year. The mean annual temperature is barely half that of the dry valley below. Rainfall is three to seven times greater—exceeding one hundred and ten inches most years.

The journey from El Rancho to Coban in the province of Alta Verapaz is like travelling from a sparse dry world to one saturated with moisture and color. In Coban many Mayan Indian communities persist with their traditional language and farming practices, some of them still wearing their rainbow-patterned cotton *traje*, or native blouses and headcloths. It is hard enough to believe that the environments of El Rancho and Coban are on the same planet, let alone within two hours' drive of one another.

This juxtaposition of environments offered an opportunity to some wealthy Spanish industrialists several years ago. Why not cut the pines out of the forests above, where perhaps only "backward" Indians might have land claims, and bring the wood back to dry in the warm sun of El Rancho? From the four kinds of pines that grow in the mountains within the ninety miles to the north of the junction, the Spaniards estimated that they could process 100,000 tons of bleached cellulose pulp a year and saw into boards another 58,000 cubic yards of timber. Asserting that they could bring in $40 million per year in pulp sales to other countries, and "save" Guatemala $20 million a year in cellulose imports for paper and other industries, the Spaniards made a deal in 1977 with former Guatemalan President Romeo Lucas García. The collaboration was called Celulosas de Guatemala, S.A., but most locals know the scheme as CELGUSA.

To build the CELGUSA pulp factory at El Rancho, the Spanish industrialists invested $92 million, Guatemalan industrialists kicked in about $20 million, and the Guatemalan government arranged to borrow another $62 million from the Spanish bank Banco de Santander. The project was boldly promoted in newspapers and its propaganda plastered on billboards. For a while, there was a nationalistic flavor to its publicity, despite the fact that foreigners controlled 46 percent of the interest in the plant. Signs erected at El Rancho proclaimed that four thousand new jobs would be created for Guatemalans. Government foresters were cautioned to cooperate unquestioningly with the Spanish entrepreneurs.

Yet there was plenty to question. According to James Nations and Daniel Komer of the Center for Human Ecology, the 4.79 million cubic yards of

pine required for the plant in its first decade of operation would have deforested as much as eighty thousand acres, about 15 percent of the remaining
land in Guatemala dominated by conifers. Up to forty thousand acres would
have been "rented" or bought outright by CELGUSA, and the anticipation
of this caused a meteoric rise in land prices in the sierra. As the pulp plant
neared completion in 1984, owners of forested land in the area were hoping
for payoffs of $500 to $1000 per acre. The only nature reserve in the region's
highlands, a twelve-hundred-acre cloud forest "biotope" for the rare Guatemalan quetzal, could not afford to buy land at that price to expand the
sanctuary's boundaries to include this bird's prime habitat.

Other environmental questions arose. The CELGUSA promotional
materials claimed that the project would create pine plantations on fifty
thousand acres of currently deforested land. These artificial forests would be
put into rotational logging during their second decade of existence, but as of
1986, no planting had begun.

If, instead, the cutting were to be done exclusively from the small remaining areas of untouched old growth and the large tracts of secondary growth,
the loss of moist tropical forest would accelerate, as it continues to do elsewhere in Central America. Guatemalans themselves conservatively estimate that eighteen million acres of forest are being logged within their country each year. If plans such as CELGUSA succeed, critics predict that
Guatemala will lose all of its forests by the year 2010.

If the land of Alta and Baja Verapaz should be deforested, the increased
storm runoff, soil erosion, and water pollution would be massive. The CEL-
GUSA plant would likely pollute the Rio Montagua and use as much as 5
percent of its water. Since that highland-derived streamflow is the source of
irrigation for Guatemala's major banana-producing region in the lowlands,
reductions in the quantity and quality of water could have severe consequences for growers downstream.

II.

In 1985, when I visited El Rancho with Howard Lyon, the acting director of
The Peace Corps volunteers in Guatemala, this dangerous scheme had al-

ready begun to deteriorate. Although wood was stockpiled in the CEL-GUSA yards, the plant had not yet opened. Rumor had it that the pulp mill's construction had begun to languish after the businessmen failed to resolve budgetary problems that plagued them for more than a year and a half.

Then, in 1986, after Vinicio Cerezo became Guatemala's first democratically-elected president since 1954, CELGUSA was asked to shut down its operation indefinitely. The new government went on record as saying that there wasn't enough wood in all of Guatemala *and* Honduras to feed a pulp mill at the scale that CELGUSA had planned. President Cerezo personally called the scheme "a national disgrace" and hoped that the closure would be permanent. Newspapers called the project "a monument of corruption and incompetence."

The decision to do an about-face on such a potentially-destructive development scheme suggests a more open political climate in Guatemala. For several years, Guatemalan foresters and conservationists had been gathering facts and clandestinely voicing their concerns to global environmental groups such as the World Wildlife Fund and the Center for Human Ecology based in Austin, Texas. Yet they had been reluctant to speak out publicly against large foreign corporate interests in their own country. They well remembered what had happened to President Jacobo Arbenz and his followers within the previous democratic Guatemalan government in the early 1950s. Some historians claim that the United Fruit Company encouraged the ousting of Arbenz after he attempted to expropriate more than two hundred thousand acres of land that the multinational corporation had cultivated.

Since the new democratic government began, conservationists have regained some confidence. They convinced decision makers that the CEL-GUSA project could have been one more of the many that are rapidly depleting Guatemala's tropical forestry resources. This one operation has been halted, or at least it is now lying dormant until economic or political changes in Central America offer it another chance. By May, 1987, the abandoned plant had caused so much anxiety among economic interests that new attempts were underway to convince the government that it could be run on reforested acreage alone. At the same time, new evidence was surfacing that suggested that extensive plantings of even-aged trees would foster the spread of a pine rust disease in the Guatemalan uplands, devastating the productivity of even more acreage.

Fortunately, neither the clearcutting nor the instant pine plantations have been reinitiated. Logs rot in the CELGUSA patios. And the green forest I saw through the clouds above El Rancho has not yet turned into red desert.

III.

Travelling through the Oriente within minutes of El Rancho, I became aware of a grayness, a spininess, a sparsity that made me feel at home. The land has the texture, color, and many of the same plants found in Mexico's Valley of Tehuacán, one of the hypothetical hearths of New World agriculture. Moreover, it shares dozens of related species having near-identical lifeforms with the Sonoran Desert and subtropical thornscrub of Sonora and Baja California Sur.

I nuzzled up to a tall columnar cactus producing delicious *pitahaya* fruit. There were old, drought-enduring friends all around me. Acacias, feather trees, and other nitrogen-fixing legumes. Limberbushes. Prickly pears. Rosettes of century plants and ground-hugging *Hechtia* bromeliads. Tree morning-glories. Shrubby yellow trumpetflowers. Brazilwood. Stout-trunked elephant trees. *Mala mujer*, an herb with stinging hairs. Pigweeds and other rainy season herbs would cover the wide spaces between perennials for just a couple of months each year.

I came during the dry season, when the barren spaces between leafless plants dominated the landscape. Soil moisture is so limited that plants must compete for that which is available, or else close up shop until the monsoons again bring rain. During the dry season, plant productivity is diminished; tree canopies are reduced in size and seldom overlap. But what is lost in tree cover is made up for by the other lifeforms that cluster with them: old man's cactus, with its tall bearded branches; viny cucurbits, rising from deep taproots; wand-like *Pereskia*, looking half-tree, half-cactus; and the spiny rosettes of the *piñuela*, analogs of the desert spoon of Southwestern deserts of the United States. The range of shapes, sizes, and styles of armor and of drought adaptations found among these plants amazes even the casual observer.

The paucity of trees within this overall diversity of lifeforms has one drawback. In the Oriente, nearly everyone uses fuelwood for cooking food, consuming about a metric ton of local timber per person each year. According to forester Peter Wotowiec, half of the families in this dry region gather all their firewood locally, while the other half buy much of their fuel, mostly from woodcutters working the outskirts of their village. The average family in the Oriente spends fifteen percent of its cash income on firewood.

Despite the abundance of trees within a two hours' drive (and most families do not own a motor vehicle), fuelwood is scarce near most villages in the Oriente. This country has long been settled, but its population is now doubling every twenty-two years. Part of this population growth is due to inmigration from other regions, perhaps because large landowners have taken over the better croplands, thereby displacing residents of those other regions. In any case, this growing population has cleared a substantial area of vegetated land in the Oriente. Fuelwood demand has grown, but to no one's surprise, the growth rates of trees in the natural vegetation have not. The semiarid pines once found in certain municipalities have been all but logged out. Nowadays, woodcutters in some villages must ride mules six to nine miles to find stands large enough to work, but the load that they can bring back on a pack animal is small. The price of fuelwood is increasing faster than that of food.

In Palo Blanco, near San Luis Jilotepeque, I walked for over an hour with Peace Corps volunteer Michael Lee to see one of several agroforestry projects that have been established to help local residents grow more wood for fuel, fenceposts, and *viga* building beams. Michael, accompanied by a parrot that seldom leaves his shoulder, showed us a small valley where the scatter of arid-adapted pines had been eliminated completely. Soon after the trees were cut, the wells dried up. Perhaps, as Michael guessed, the watershed had lost its capacity to absorb rainfall, most of which now runs off rather than replenishing the local aquifer. He had therefore suggested to some local residents that if they would consider planting rows of trees in their cornfields, he would provide the nurserystock for them.

Their reply made Michael realize the severity of the problem. They asked him to plant entire fields in trees that would produce tall, straight lumber and fast-growing fuelwood, leaving no room at all for crops. They had other fields, but here the timber could prove more valuable than their maize plantings. And what good was corn if there was not enough wood to cook it, and not enough lumber for a good kitchen roof?

IV.

Stopping CELGUSA's destruction of tropical moist forests is infinitely easier than curbing the diffuse depletion of woody plants in the more arid tropical zones of Guatemala, or in any other dry land for that matter. You cannot win the war simply by winning one battle. Human population growth, and the pervasive environmental degradation associated with it in the arid subtropics, amounts to ten thousand miniature battles going on every day. There is no sinister industrialist to target as the enemy. There are just individuals like you and me, many of whom merely want what we take for granted: a leakproof roof above their heads, enough food for their families, and enough fuel to cook with.

And yet these modest needs, when multiplied, put pressure on a land where productivity is severely limited by the scarcity of water. There is only so much resilience in these plant communities before they take on the telltale signs of desertification. Take away the moisture-holding vegetation and the rich organic pockets of soils where tree canopies cast their shadows, and degraded dry tropics look much like true deserts.

The degradation of dry forests has, until recently, not been much noticed; the thornscrub itself has gone virtually unprotected. A *Science* magazine editorial in 1986 noted that "Tropical dry forest, which once accounted for over half of all tropical forest, has received little attention internationally amid the alarm over the demise of rainforests. One reason is that there is so little of it left." Ecologist Dan Janzen contends that so much of the tropical dry forest has been lost that this vegetation type, not the rainforest, should be considered the most threatened of the major tropical forest types.

When dry forests are alive and thriving, they often go unnoticed. And when wiped off the map, they might inadvertently end up in two sets of mortality statistics. Rainforest conservationists sometimes lump them with other losses to estimate rates of tropical deforestation. And desert biologists use the degradation of dry forests at the edge of the tropics as examples of desertification, for these lands dry up both biologically and hydrologically. In global summaries of vegetation destruction, there is the chance that each dry forest obituary is being counted twice, after submission from different experts.

How much of Central America was once covered with dry forest? Ecologist Dan Janzen has offered a graphic answer: "When the Spaniards first hit the Americas in the sixteenth century, there was dry forest over an area the size of France, or five Guatemalas."

In Costa Rica, where Janzen is valiantly working to set up parks and restore degraded vegetation to its former richness, the drought-deciduous forest area has shrunk in two decades from 20 percent of the country to 2 percent. Worse yet, only .09 percent is currently protected in land reserves.

But naturally dry does not imply naturally poor. Dry tropical forests have about as many insects and mammals as rainforests, if Janzen's estimates are correct. There are 20 to 40 percent fewer bird and plant species in dry forest than in rainforest. According to other noted ecologists, Robert May and Thomas Givnish, many more plant growthforms exist in the more arid, open vegetation types than in rainforests. Moist tropical forests are dominated by tall-trunked woody plants that place their fruits or seeds high above the ground before letting them loose for dispersal. Throw in epiphytes and viny lianas, and rainforests still do not approach the architectural free-for-all found in certain drier kinds of vegetation.

As aridity increases in the tropical belt, so does the number of plant architectural strategies represented. Where moisture becomes both scarce and unpredictable, no single lifeform works best all the time. Over twenty lifeforms can be found intermixed in the arid subtropics, from columnar cacti and palms, to dwarf shrubs and leaf succulents, to root parasites, short-lived bellyflowers, and rock-mimicking stem succulents.

Writing in *Science* magazine, Constance Holden relates this lifeform diversity to the habitat variety characteristic of drier tropical settings: "Dry forest is in some ways more complex because of the extreme weather changes and the . . . variability in rainfall, which make for a greater variety of small habitats."

Considering the tremendous habitat heterogeneity where dry subtropical forests grade into desert thornscrub, it is no wonder that variability both within and between plant species is high. Geneticist José Esquinas-Alcazar calls these intermediate arid transition zones "the laboratories in which many new adaptive complexes of plant groups are produced." During his FAO-sponsored work collecting and conserving crop genetic resources in Latin America, Dr. Esquinas has observed that these evolutionary laboratories are rich in wild progenitors and relatives of food crops.

Scan the list that Esquinas has roughed out and you will find drought-tolerant wild plants that are kin to many New World crops. Many of them are from the arid subtropics or dry tropics. Wild tomatoes. Several cassava or manioc species. The tequila agave, and its fiber-producing kin. Wild gourds. Teparies and other beans. Two potatoes, the "papa amarga" of South America, and "papa guera" of Mexico. Peanuts, amaranths, prickly pears, cottons, and grapes are just a few of the others.

Intermediate arid zones may also be where seed agriculture actually began in the Western Hemisphere. Some of the oldest agricultural remains in the Americas come from the rainshadowed Tehuacán Valley in southern Mexico, where vegetation has much the same texture as that of Guatemala's Oriente. Farther north, the Ocampo caves in Tamaulipas are nestled within a fairly abrupt gradient from cloud forest to Chihuahuan Desert. Both are sites of domesticated plant fragments that are thousands of years old.

It makes sense that early agricultural experimentation might have emerged in such a setting. As prehistoric human populations grew, there was out-migration from sites where water and fruits were available year-round, to more marginal areas where aridity dominated much of the year. Wouldn't these be likely areas for attempts to intensify their food procurement during the wet season in order to have enough on hand during dry periods? On drier, more marginally productive wildlands, many annual seed plants that people had once known in wetter habitats would still germinate and produce a useful harvest if sown when the rainy season began. The dry tropics, not the rainforests or the dune-covered true deserts, are thus the most probable wellspring of seed agriculture in the Americas.

That is why the draining of plant gene pools in drought-deciduous tropical forest and thornscrub is so tragic. New World farming cultures have found these plant communities to be their sources of food crops for the last ten millennia. While Norman Myers and Catherine Caulfield may be correct in calling the tropical rainforest *the* primary source of medicines on earth, it can be argued that the floras of drier areas have just as generously served as the founding gene pools of most food crops.

Nevertheless, this is a connection that is easily obscured as useful plants have migrated in space, through time. The Hopi Indian farmer, planting a half-dozen lima bean varieties in his sand dune field in semi-arid Arizona, does not know that the northernmost wild limas twine into drought-deciduous shrubs on the Tropic of Cancer in Mexico. He may not ever need their genes, but the commercial lima grower in the dry valleys of California may need to draw upon them at some point in the near future.

The trouble is that plant breeders, farmers, and consumers have always treated this primary source as if it will always be "out there," in reserve as raw material for the food production system. As John Creech of the U.S. Department of Agriculture observed a few years ago, "The attitude of seed conservation has really been one of reaction. If we need rare strains to breed a stronger variety of grain in the event of an epidemic, we go out and collect them. We have felt that we can always go to the country of origin and get the seeds we want."

Those times are passing all too quickly. The current rate of vegetation destruction is certainly diminishing the accessibility of wild plant resources. As much as 75,000 square miles of tropical vegetation types are being deforested or converted each year in such a way that plant extinctions are likely to occur on that land. Even if that deforestation rate slows, the outlook for global species diversity is bleak unless additional lands are protected. And because traditional farming and gathering peoples of the Neotropics are being acculturated with the same speed that their forests are being cut, ethnobotanist Michael Balick predicts that there is time enough for only one more generation of scientific research about the rarer tropical plants now used by these peoples before the plants and the folk knowledge about them are both gone. Conservation biologist Daniel Simberloff has made another prediction: "If tropical forests in the New World were reduced to those currently protected as parks and refuges, by the end of the century about 66 percent of all plant species and 14 percent of [plant] families would disappear. . . . The imminent catastrophe in tropical forests *is* commensurate with all the great mass extinctions except for that at the end of the Permian period."

If these extinctions were only occurring in the wetter tropics, we would be in enough hot water. Yet drier regions are also being converted and degraded on a massive scale, and they too are rich in *endemics*, or species of plants restricted to small areas. The United Nations Environment Programme estimates that 2.4 million acres of semi-arid or subhumid land is annually reduced to desert-like conditions. Within true, hot deserts, another 8.5 million acres is depleted of its cover, and left as barren sands. As a result, we are witnessing the genetic wipeout of entire species of wild cassavas, prickly pears, sunflowers, and agaves.

In a matter of years, a newly discovered plant population can be eliminated before it is studied, appreciated, or collected. I recall the story of Daniel Debouck, now a field collector for the International Board of Plant Genetic Resources. In the 1970s, he encountered an interesting wild bean population in Durango, Mexico. At the time of his visit, no seeds were ripe for collecting. He returned to it just a few years later, but was unable to find a single plant. Increased pressure from livestock had eliminated the population altogether.

V.

There is another bean population, of wild limas, first recorded decades ago by Mexican botanists, that I had always wanted to see. It is the one that lies smack-dab on the Tropic of Cancer, as far north as wild lima beans go. As I headed south from the Sonoran Desert with friends, I wondered if those bean vines had persisted over the intervening years, surviving the growing intrusion of people and cattle into its habitat.

We passed from the creosotebush flats of Arizona into the complex mix of tall cacti and mesquite that characterizes much of Sonora. Gradually, as we drove south, we could see the effects of greater summer precipitation and fewer freezes. The vegetation became more dense. Trees, still drought-deciduous, coexisted with many other lifeforms, including a few epiphytes that hung from their branches. Leaving Sonora for subtropical Sinaloa, old standbys like the saguaro cactus disappeared, and tree canopies became umbrella-like in the thornscrub.

As we pushed into wetter, more southerly climes, the number of coexisting tree species increased dramatically. Trees grew taller, too, and more varied in shape. They were fighting for light, as water became more available and the vegetation consequently became more tightly packed. The remaining cacti were enormous—much taller than saguaros—and were covered with a variety of epiphytes just as the trees were.

We began the ascent up to the Continental Divide close to the Tropic of Cancer, not far from Mazatlán, Sinaloa. As we rose from sea level into the sierra, we hit a transition where a number of trees began wearing leaves, while others remained stark naked.

There, along a stream draining from the mountains, I found the lima beans. Flowering in February, climbing four yards up into riparian shrubs at the floodplain's edge, their lovely flowers dazzled the local bees. And the foreign botanists. The beans had somehow survived the passage of time.

We continued up toward the Devil's Spine on the continental backbone. Cacti and thorny shrubs disappeared from the vegetation; it became so green my eyes hurt. There, in a cloud-moistened, frost-free perennial forest, tropical fruits were perched high in the canopies. Fog rolled in on us and darkness gathered. Unable to find much horizontal ground, we set our tents up on the edge of a dirt road leading to a sawmill. Tired, and only half-aware of our immediate surroundings, we climbed into the tents and went to sleep.

At dawn, I stepped out of my tent and nearly dropped a hundred feet into someone's slash-and-burn maize *milpa* below. Instead, I walked up a hill crest to the west, to look down upon the tropics where woodfires in a few pueblos were glowing like embers in the perennially green canopies. Below these tropical villages, the drier forests slope toward the Pacific coast. They share many species with Costa Rica's Guanacaste National Park, where Dan Janzen has been working for decades; with El Rancho, in Guatemala's rain-shadowed valley; and with the tropical edge of the Sonoran Desert around

Alamos, Sonora, where I had collected wild relatives of crops among Guar-ijio and Mayo Indians.

I sat above a cliff and turned to watch the sun come over the continental divide, showering the lands to the west with light. It was then that I prayed for the fruits that the tropics have offered us: the figs and custard apples; avo-cados and guavas; vanillas and chiles. And I prayed that the sawmills and woodstoves around me would spare enough of the trees so that my children can someday see dawn come to an untouched forest on the Tropic of Cancer.

CHAPTER THREE

Fields Infused
with Wildness

I.

Flocks of birds were foraging in the recently-harvested fields, gleaning the ears that had been too small to gather, the pods that had been too green to pluck from the vines before the frost arrived. The Tepehuan Indian fields were scattered across the small Chihuahuan valley of Nabogame, not detracting from the mountain habitat so much as fitting modestly within it.

These fields, fenced as patches on the sloping grassy tongue of the valley, were not manicured environments; not merely corn, nor beans, nor squashes planted by human hands. The croplands held within their reaches a wealth of wildlings, keeping the fields in a state of flux: a weedy maize named teosinte; a flutter of ravens; a weaving of wild bean vines; a heap of husk tomatoes. The sierras threw their shadows down across the Indian fields, the unruly-looking crops, and the semi-tame turkeys that meandered among the stalks.

Here and there, modest pockets of tended land dot this stretch of the Sierra Madre, but the overall impression is of wildness as a pervasive force. A spontaneity is expressed even within the cultigens that elsewhere seem so uniform and predictable, say, in the rolling stretches of the Corn Belt. *Milpa* here does not mean "cornfield" in the sense of the kinds found in Iowa, where double-crossed feed corn hybrids can extend in curvilinear rows arching from horizon to horizon.

In Nabogame cornbins, I saw a turbulence of kernel colors and textures: purples, whites and blues, flours and flints intermixed. Endowed perhaps with the ancient wisdom of teosinte's genes, the Indian corn has developed considerable hardiness to cold and drought. No matter how good the farmer who plants them, these crops must still face the freezes, hails, and thundershowers that frequent their montane setting. The acres of plantings, replete with weeds and birds, will always be dwarfed by the rugged mountains above them.

I could not get to this place directly from my home, for there are no straight highways between the two. Instead, I had to zigzag in upon it, like a hummingbird trying to reach a flower containing some essential nectar. From where I work in Phoenix, the linear distance to Nabogame, Chihuahua, is no greater than that to Denver or San Francisco. Yet it is more than the lack of a direct route that makes the mixed fields there seem light-years away. The fields of Nabogame, Chihuahua, express a raw richness and variability not generally found in agriculture today; not even all Indian or all Tepehuan fields are as diverse and wild as these. The farming village of Nabogame embodies a set of cultural values not to be found on a downtown corner in Denver, Phoenix, San Francisco, or even Farmington. In Nabogame, generation after generation of plants, and of animals, and of local knowledge about them, have formed an unbroken chain across the ages.

But what can one learn of such a persistent tradition in a few days in Tepehuan Indian country? I travelled there only after spending considerable time listening to relatives of the Tepehuan talk of their crops and the wildness that surrounds and sometimes permeates them. It seemed to me that many people of Tepiman heritage had not forgotten, even in their agricultural pursuits, the connection between wildness and wellness. I remembered that for some of the Tepiman-speaking groups, such as the River Pima and Tohono O'odham, the words for "wildness" (*doajk*), for "wholeness" and for "health" (*doajig*), appear to be etymologically related. These words are probably rooted in the term *doa*, "to be alive" or "to be cured," as is *doajkam*, "wild, unbroken beings." Pima and Tepehuan people also have ways of describing the introgression of wild or weedy genes into a cultivated crop, and a sophisticated sense of the relationships between domesticated and spontaneous species.

I was as much interested in how the Tepehuan people discussed the relationship between corn and teosinte as I was in seeing these plants growing side by side. Like friends of mine before me—Garrison Wilkes, Howard Scott Gentry, Mahina Drees, and Barney Burns—who had all visited Nabogame over the years, I sought to collect extras of seeds that farmers could spare. But I also felt that the folk knowledge about the seedstocks was as important as conserving the seeds themselves. I hoped that I could use my rudimentary understanding of Tepiman languages to see how such knowledge was encoded in Tepehuan culture.

My route to Nabogame was circuitous by necessity. I took two planes and a bus before I met my friends Laura Merrick and Salvador Montes in Hidalgo de Parral, Chihuahua, one of the entry points into the sierra. Salvador, a breeder of native vegetables for the Mexican government, and Laura, a conservation biologist specializing in squashes and gourds, were collecting seeds from squashes and gourds for the Food and Agriculture Organization of the United Nations. I was collecting wild beans for the same. But we cast our nets far wider than necessary to catch only these targeted species, since we knew that scientists' trips into this part of the sierras were few and far between. We drove a day beyond Hidalgo de Parral, mostly on dirt roads winding through pine-covered barrancas, before roads vanished completely.

In a village of *mestizos* on a windswept mesa, we searched for mules or horses to rent for the last leg of our pilgrimage. Before the next dawn, we set off riding on pack animals to cross several wooded canyons and hilly ranges, and reached Nabogame on the Rio de Loera by noon. There, a few Tepehuan and *mestizo* families welcomed us, as they did a travelling band of Tarahumara youths who had come visiting from their ranchos to the north.

After the long ride it was a pleasure to stand on the ground again, talking with the Nabogame families, sharing drinks and stories with them, exchanging native garden seeds. After a while, however, my eyes began to wander to the fields and mountains beyond the homesteads' dooryard gardens.

"Could we see where *maizillo* grows?" I asked a lady named Clorinda, who had been hostess to my friend Garrison Wilkes two decades before, when he had studied this "little corn."

"I'll show you, it's close, not far behind us," she said, pulling a sweater over her shoulders to shield her small body from the wind. "We've harvested the *maiz* already, but the *maizillo* will still be standing."

Clorinda beckoned us to follow her past a small patch of runner beans climbing up cornstalks, to another, larger field of maize. She held her hand like a visor above her eyes, for the sun was near the mountaintops, backlighting the dried stalks that rattled in the wind. Still, her eyes moved amidst the stalks until she noticed one with a diminutive husk still attached to it.

"There's the *maizillo*," she said, pointing. Then, gesturing a few yards behind us, "There's another. You will begin to see them. There are many."

I sidled up to a stalk she had pointed out, although at first it appeared not much different in size or color from the stalks of the flint corns that dominated the field. But when I dehusked the lateral inflorescence, it held not a heavy cob, but a spindly rachis with a few angular seeds not much bigger than wheat grains.

The little spike of hard-edged seed is what the Tepehuan call *kokoñi ushidi*, "raven's planting." Perhaps as the legendary Coyote had done among the crops of their Piman relatives, ravens had planted a grain in Tepehuan fields that resembled corn, but had a weedy, untamed bent to it. It may be that ravens are tricksters among the Tepehuan like Coyote and Raven are in the folklore of many tribes. In any case, ravens still frequent the fields of the sierra, and the *maizillo* continues to grow wild in the cornfields.

Call it *maizillo*, or *kokoñi ushidi*. Call it, as scientists have in decades past, *Euchlaena mexicana*, or as they do today, the Nabogame race of *Zea mays* subspecies *mexicana*. Refer to it by the widespread Mexican term, *teosinte*, from the Nahuatl "mother of corn." Whatever name is given to it, this Nabogame population of weedy corn is the northernmost known to science. This and other small populations in Tepehuan fields and nearby washes are separated from the core area of Mesoamerican teosintes by at least three hundred miles. Curiously, the closest teosinte site to the south is near Durango City, where the Tepehuan of Nabogame and neighboring ranchos are said to have come from, following their rebellion against Spaniards around 1616.

However long it has been isolated from other teosintes, the wild maize of Nabogame has become distinct in terms of a number of genetic criteria. Its chromosome knobs, its enzyme "fingerprints," and its mitochondrial DNA all exhibit minor differences from other annual and perennial teosintes.

Clorinda and her relatives had granted Garrison Wilkes permission to collect seed from their fields about twenty years before my own visit. For many years, his seed harvest from Nabogame valley was the only example of this teosinte race kept in many of the gene banks around the world. As these germplasm repositories sent out samples of this seedstock over the years, their supplies dwindled faster than their growouts could replenish them. Garrison requested that we ask Clorinda if the seed could be collected again for gene banks in Mexico, and she saw no trouble with that. She had cherished and still kept letters that Garrison had sent her over the years.

As Salvador, Laura, and I stuffed each single plant's yield into a different envelope, hoping to capture a sampler of the variation within the field for later analysis by geneticist John Doebley, I recalled the words of explorer Carl Lumholtz, who first stumbled upon this teosinte at the turn of the century: "Around Nabogame grows a plant named *maizillo* . . . more slender than an ordinary corn plant and the ears very small. . . . However, several Mexicans assured me that, when cultivated, the ears develop. After three years they grow considerably larger and may be used as food. . . . I was told that people from the Hot Country come to gather it, each taking away about one almud to mix with their seed corn. The combination is said to give splendid results in fertile soil."

Although they have been grown together for hundreds of years, Nabogame maize and the little teosinte do not completely integrate with one another; there may be some gene flow between them, but one is not overwhelmed by the other. John Doebley underscored this point to me after he had analyzed the samples that Laura, Salvador, and I had sent to him: "Despite the fact that the teosinte and maize are from the same field, they maintain themselves as separate entities. There is some leakage between them . . . a few alleles for which I could make a case for introgression . . . but most remain very different."

John had grown out seedlings of our collections in a greenhouse, and used extracts from them to analyze their enzymatic similarities. Recently, he showed me his results: banding patterns stained on gel-covered paper, which indicated how particular enzymes had migrated up the gels during a process called *electrophoresis*. These "plant fingerprints" indicate a few enzymes that may have flowed from teosinte into maize in the same field, and one case where an enzyme found in teosinte may have come out of maize. John interpreted these cautiously, for there the Nabogame maize shares more enzymes with other Tepehuan Indian maizes (grown away from teosinte) than it does with teosinte from its same field. Nevertheless, the maize-teosinte gene exchange which Lumholtz first conjectured about probably occurs at a low frequency, and it is this crop/weed introgression that has attracted attention to Nabogame over the decades.

In response to this attention, Nabogame farmers have shared their teosinte with farmers from neighboring villages, with Tepehuan families from the more tropical barrancas, with germplasm collectors from places as distant as Mexico City, Boston, and Beltsville, Maryland, and with the local ravens. The teosinte genes flowing up through their fields are a wellspring where visitors come to obtain a refreshing drink. Yet Nabogame villagers have not allowed their teosinte reserves to be depleted, even though the plants are sometimes weeded, sometimes grazed, and periodically collected. When many of the surrounding villages began growing illegal drug crops in the 1960s, Nabogame farmers nevertheless continued to grow maize and teosinte.

Garrison Wilkes visited the sierra in the sixties, staying one night at an inn where everyone else worked at cleaning drug plant harvests while he contemplated more mundane matters such as corn's history and diversity. But when he arrived on the Rio de Loera, he was reassured that "wild teosinte is abundant in the region, and hybrids almost as abundant in the fields as on the margin of the fields. . . . [T]he method of cultivation of maize has not changed appreciably in the last hundred years."

Garrison Wilkes observed people harvesting hybrids between Indian corns and weedy teosinte. He was told that farmers from other areas would bring in their corn seed, so that it might be exposed to teosinte by being grown in fields where the latter was abundant. The resulting seed, when planted, was said to produce flintier kernels and greater yields. Although the teosinte gene flow continues in very few domesticated corn varieties in the Americas, it has a small but significant influence on four races of Mexican corn that are still grown by the Tepehuan: Apachito, Chalqueño, Conico Norteño, and Harinoso de Ocho.

Twenty years after Garrison Wilkes's visit, his hosts not only remembered him, but passed on essentially the same message to me: teosinte was tolerated and protected, for it invigorated their corn. They are not aware that Mexican botanists have listed their teosinte as "potentially vulnerable" in a country-wide survey of rare and endangered plants. Nor do they know that isolated gene pools of crop relatives are now of international concern, or that geneticists are eager to freeze the seeds from such populations to "save" them. They simply do what their families have done for hundreds of years; they keep teosinte alive in their cornfields, alive in their hearts.

II.

Halfway between Nabogame and Phoenix, Juan Ignacio Humar knows that wild chiles are alive just past the margins of his cultivated chile plots, and that they keep heartburn alive in his community. From what I could figure, tiny halictid bees must carry pollen from wild chiles thirty yards away from his field to his well-tended, plump-fruited plants. From what Juan told me, the chiles from the resulting plants are so pungent that some of his friends can hardly eat them.

"The wild chiles of the countryside inoculate those in my garden with their piquancy," Juan said, somewhat understating the case, considering the pain those natural hybrids have brought to his friends.

"Well," I asked, "does that injection of heat make them bad?"

"Oh, no. I eat the chiles, hot or not. There's great variability in the spiciness of cultivated chiles around here, especially if the wild ones have inoculated them. I never know what one will be like until I bite into it."

"When you get a lot of the hot ones, can you sell them?" I asked, wondering if Juan may have a viable hybrid variety on his hands. "Do you save the seed?"

"Well, I save the seed because I like eating them enough to plant them again. It's my own seed, not a selected commercial variety. But I believe they're too hot to sell to the buyers who come around for the market in the city . . . I simply keep them for my own pleasure."

Juan lives on the subtropical edge of the Sonoran Desert, in an ancient Pima village where very few residents now speak the indigenous language. Nonetheless, some of their fields have a healthy mix of crops and wild plants in them. They weed some of the latter out, but weed seeds are periodically replenished when floodwaters from the Rio Yaqui and its sidestreams rise high enough to spill over into the fields. Certain wild plants and animals are harvested as much from these fields as they are from uncultivated habitats in the hills nearby.

Piman-speaking farmers once dominated the agricultural production of the Sonoran Desert, from the Lowland Pima in Onavas, Sonora, and the Sand Papago at Quitovac, to the Tohono O'odham west of Tucson and the Gila River Pima near Phoenix. Over the years, when visiting these people's fields, Amadeo Rea and I have kept running lists of the wild plants we see in them. Oftentimes, we'd press specimens of plants whose names we were unsure of to enable later identification. Overall, we've encountered more than 183 species of wild and weedy plants in the field communities tended by Piman-speakers in the Sonoran Desert. Even though only one hundred acres of traditional fields of Tohono O'odham farmers remain in crops, over 110 bird species and 130 wild plant genera can be found around them.

Ask Pima farmers why they let volunteering plants persist among their intentional sowings, and they will give you a potpourri of answers. One plant may have edible greens, while another is host to edible insect larvae. Cacti and thorny shrubs emerging at the field margins can be shaped into a hedge that deters livestock from the plantings. Annuals may be allowed to shade the ground or certain seedlings for a while, but are then cut and used as mulch or compost to fill in spots scoured out by floodwaters that moved across the field earlier. Trees like mesquite are left to shade tired workers from the desert heat, and their leaf litter is said to make the surrounding soil richer.

From such responses, it would seem that Pima or O'odham only value the wild plants that directly benefit them. They suspect many benefits that might escape our eyes, from gene enrichment of crops and nutritional enrichment of their diets to soil enrichment of their fields. But those are the easy answers, the kind that I heard the first time I walked with a somewhat shy farmer through his field. If he knew that I was interested in useful plants, he would give me answers that satisfied my questions.

Sooner or later, I noticed that other wild plants and animals were left alone around fields even though they gave farmers no "economic" benefits, monetary or otherwise. Once I discovered some gopher holes in a friend's field, near one of his melon vines, which appeared damaged. I wondered if he controlled these burrowing animals.

"I thought about it, now that my wife is old, but I still leave those gophers alone."

"Now that your wife is old?"

"They say that if you mess with the gophers it will disrupt the ladies' monthly sickness when they are young. You aren't supposed to kill those gophers or even touch the plants they've been chewing on. You just let them be." Then he smiled. "Now that she's too old for that, I've been thinking about getting some of that . . . what do you call it? Gopher-Go? But I just don't know."

The moon may not call to the man's wife anymore, but in a nearby village, some plants are still called upon once a year, as a phase in the annual cycle returns. A yellow-flowered composite shrub is one of more than one-hundred-and-twenty wild plants on ten acres of the Quitovac oasis, where for decades Mexican O'odham have used it as a decoration in a summer harvest ceremony. I have found other, rarer plants there as well, such as an agave that produces little plantlets atop its flowerstalk and a mutant hedgehog cactus with an albino blossom.

In the oldest, most stable of these Native American farming villages, plant species that are otherwise rare or at the limits of their ranges have gradually accumulated. Mixed crop fields or fallow ones, hedges, ditches, dooryard gardens, and pastures as well as ungrazed scrublands—all offer habitats to an astonishing variety of plants and animals. As Jan Alcorn has shown among the Huastec Maya, the spatial heterogeneity of habitats created by native land management fosters both intraspecific and interspecific diversity.

Some scholars have argued that this diversity lends stability to both wild and cultivated ecosystems, but it may be the other way around. David Ehrenfeld suggests that overall biotic diversity in small farming communities is enriched and conserved as a result of their cultural stability through time. In *The Last Extinction*, ecologist Ehrenfeld argues that it is in such communities as this that "conservation becomes reality, when people who are not actively trying to be conservationists play and work in a way that is compatible with other native species of the region. When that happens—and it happens more than you think—the presence of people may enhance the species richness of the area, rather than exert the negative effect that is more familiar to us."

III.

Those all-too-familiar negative effects of our species are rapidly transforming the face of the earth and are pervading much of the modern world. Ironically, some ecologists argue that the destruction of biotic diversity was as characteristic to preindustrial societies as it is to industrial societies. If they assume that destructiveness is the human norm, it is easier for them to write off our failures.

In a widely quoted editorial in *Nature* magazine, Jared Diamond contended that prehistoric cultures did as much environmental damage and caused as many extinctions as have their modern counterparts. Diamond, a Los Angeles biologist and expert on island bird extinctions, claims that it is an "environmentalist myth" to portray any preindustrial peoples as "gentle conservationists." He apparently does not agree with historians such as Calvin Martin who have argued that "on the whole, the North American Indian earns high marks for his cautious uses of plant resources. . . . [Indian cultures proceeded] cautiously because . . . Nature could strike back against abuse."

Diamond may have been trying to oppose the simplistic notions that "noble savages" were in static balance with nature, and that "primitive" religious beliefs always translate into ecologically-sound land management practices. I understand his impatience with such romantic notions, which serve to idealize prehistoric practices. Diamond dwells upon some extreme but authentic examples of biotic communities being ravaged by our pre-Columbian predecessors. Yet in citing only the ancient extinctions of island faunas, and the late prehistoric deforestation of Chaco Canyon, Diamond sloppily subsumes paleolithic hunters, small-scale farmers, and city-state traders into his catchall term, "preindustrialists," without regard to the differences between various prehistoric ways of making a living and knowing this world.

In fact, the ecological impacts of small-scale farming traditions are not among those specifically called on the carpet. At one extreme, Diamond cites the quick demise of a docile fauna when pioneering hunters invade islands, new territories in which they have not yet learned any constraints. At the other extreme, Chaco Canyon deforestation was driven by the use of one hundred thousand conifers for the multi-storied pueblos, an Anasazi city-state. Like the shapers and defenders of industrial economies, the priestly elite of Chaco were preoccupied with extra-regional trade of commodities rather than with local self-sufficiency. They had left behind the minds of subsistence farmers centuries before.

Julio Betancourt has hypothesized that Chaco's farmlands were abandoned when erosion followed the cutting of lumber and fuelwood in the watershed above the pueblos. At first, the loss of local natural resources may have intensified the pueblo's dependence on commodity imports. Soon, economic instabilities must have toppled the puebloan power structure. Although the Anasazi exodus from Chaco Canyon occurred eight hundred years ago, Betancourt believes the vegetation there has not yet recovered.

But there is a middle way between the economic strategies of invading hunters and those of city-state traders. This is the way followed by farmers who continue to till patches of soil amid mountains and in remote valleys in many places around the planet. In his probings of folk biological knowledge, Cecil Brown has compared farmers who hunt, gather, and tend small fields with other kinds of people, such as non-agricultural hunter-gatherers, factory farmers, and their urban analogs. The mixed subsistence farming cultures, Brown discovered, have a richer knowledge of wild flora and fauna than do cultures dominated by any other form of livelihood. On the average, the number of plants and animals known and named by small-scale farmers is about three times that named by either hunting bands or by city street gangs.

Oftentimes, sowers of small plots in marginal lands must also practice hunting and gathering. This keeps them keenly interested in the identity and availability of wild resources. Living at higher population densities than those of free-ranging hunters, such farmers must rely on wild resources in years of drought, crop failure, and famine. Widely-dispersed nomads can afford to ignore localized wild resources so long as they can move off to other areas where the effects of drought and other climatic extremes are not as marked. Tied to particular plots of land, village farmers manage their patches in ways that foster a heterogeneity of habitats, broadening the variety of wild organisms.

Brown has examined agrarian cultures around the world that can name from a thousand to twenty-three hundred different plants and animals within their neighborhoods. He believes that these high numbers are not merely a result of the farmers' selecting and naming so many varieties of cultivated seeds and livestock breeds. Brown maintains that peoples who mix farming with gathering and hunting have retained a rich knowledge and appreciation of wild diversity.

Other observers of peasant fields and their uncultivated surrounds suggest that this detailed knowledge translates into both conservation and wise utilization of plant genetic resources, domesticated or otherwise. In *Conservation Biology*, Miguel Altieri, Kat Anderson, and Laura Merrick argue that traditional farmers realize that the source of their well-being resides as much in the untended margins of their fields as in the rows of crops they have sown. These are not the stereotypical "gentle conservationists" whose continued existence Diamond doubts, for they harvest and hunt, burn and boil many of the organisms that are maintained by Indian land management practices. Their intimacy with the plant and animal populations around them alerts them to minor declines, so that land use or harvesting pressure can be relaxed if the species means much to them. Legends, hunting taboos, and other cultural encodings may constrain their behavior so that frequently utilized species in distress are offered enough space and time to recover. It is a pragmatic kind of conservation, though perhaps an ethnocentric one, yet it has worked longer than any modern conservation programs.

IV.

When farmers intensify or expand their cultivated domain, however, their feel for the wild world may diminish. Intensive agriculture pushes many wild organisms out of the way. Once the wild thread has been pulled out of their fields, the texture that supports hunting and gathering unravels. When farmers chop down their hedges in order to plow clear up to the road's edge, the wild creatures flee or perish.

"Intensive monocropping of high-yielding varieties leads to the biological simplification of the environment," Cecil Brown writes, and to a "loss of intimacy with the world of plants and animals." His studies indicate that most workers on mechanized farms are little different from city dwellers when it comes to firsthand contact with a wealth of wild species. Their folk taxonomies, jokes, and legends are largely deforested; few beasts roam their dreams.

Some fields persisted as small pockets within wildlands for centuries, and life flourished around them. Steve Emslie has studied the faunal remains among New Mexico pueblos that, unlike urban Chaco Canyon, did not "get big." He has found seventy species of birds among the bones left in the ground at these sites, from A.D. 1250 to the present decade. Many of the avian species recorded at these sites do not normally occur in any abundance on the Upper Rio Grande or its tributaries today.

In attempting to explain what nurtured this diversity, Steve Emslie points to the seedheads, the old *acequia* canal systems carrying water, and the insects carried along by both. Such features served as magnets for martins, warblers, flickers, corvids, and swallows. These species increased in abundance as a result of human-managed habitat heterogeneity. Other birds, altogether unusual for the Upper Rio Grande, also appeared there. Harlequin Quail, Wood Ibis, Black-billed Magpies, and Boreal Owls lived in these farming villages, far from where their core ranges are now.

Birds flocked to these riverside fields and wild plants proliferated there. Archaeologist Joseph Winter has recorded more than one hundred species of wild plants historically managed by the puebloan peoples and another fifteen species that were encouraged by semi-cultivation. He then searched back in time, to demonstrate that at least two dozen of these species were also significant resources around prehistoric pueblos such as Hovenweep on Cajon Mesa, in Walnut Canyon, and on tributaries of the Rio Puerco. These attractions between plants and humans are not skin-deep; they are complex bonds that develop over centuries of time.

V.

I remember a meeting I had once with the foremost authority on squashes and gourds, Thomas Whitaker, and a plant conservation policy expert, David Kafton, Director of the National Council on Gene Resources. Kafton had come upon the idea of setting up *in situ* reserves for rare wild relatives of squashes, and came to us to obtain suggestions on where to locate these reserves.

After elaborately explaining his theory of why park reserves might be more useful than placing seeds in frozen storage within gene banks, Kafton must have noticed that the elderly Dr. Whitaker looked a little impatient. Kafton finally thought he would use a tangible example to capture Dr. Whitaker's imagination.

"Dr. Whitaker, can you name a pristine place in Mexico where an endangered gourd still grows, one where a nature reserve could be set up for it?"

"I don't believe there is such a place," Dr. Whitaker replied quietly.

"What? Surely there must be some relatively pristine places left in Mexico?" Kafton had found Whitaker's answer incredible.

"There are," I recall Whitaker answering, "but they're not the places where the endangered gourds grow. The rarest wild *Cucurbita* in Mexico are found in one or two hedgerows, or on the edge of a little dooryard garden, twining up some lady's clothesline. I'm afraid they're not in natural vegetation. They've been associated with the Mexican people for thousands of years!"

A number of persistent traditions on this continent have harbored a diversity of plant rarities within them, with little concern for whether these organisms are purely natural or culturally derived. These traditions are now exposed to unprecedented social and economic pressures that could easily destroy them. Culture change has always happened, but its current rapidity is now disrupting ancient communities where both wild and cultivated plants have been conserved together for centuries. These plants are being separated from one another as their cultural habitats are being torn apart. Something that has long kept our cultigens and even our peopled landscapes healthy and tolerable is now disappearing. That valuable entity is wildness. If it is lost from the world around us, we will lose something within ourselves as well.

CHAPTER FOUR

Invisible Erosion:

The Rise and Fall
of Native Farming

I.

We're beyond the end of a long runway, at the bottom of a dry irrigation ditch. Just a few hundred yards away, a jet lifts up off the ground and thunders over our heads. My three-year-old daughter's hands fly up to cover her ears, and she hunches over in fear. Horrified by the deafening noise of the aircraft above her, she looks around for shelter. But the ditch bottom is largely barren clay, dried and cracked. The few saltbushes nearby are whipped violently by the tailwinds of the jet. Failing to find any cover that will muffle the roar, she cries out to be held, and runs toward me down the ditch.

Fish as large as my daughter once swam down this prehistoric Hohokam canal. The canal itself is among the largest built in pre-Columbian North America. Among the fish it formerly carried was *Xyrauchen texanus*, the humpback sucker, reaching up to a yard in length and weighing thirteen pounds. Humpbacks were once common throughout the Salt and Gila watersheds in central Arizona; they frequented the main ditches of the River Pima Indians, who came after the Hohokam. Today, elderly Pima men can still describe with some flair this fish, which they call *o'omuni*: "It has a yellow belly [and] looks like trout, more or less, but the back is high, the tail thin, and it comes down like this . . . " they say, shaping with their hands the low spot between the hump and the tail. For Sylvester Matthias, its actions are also well remembered: "*O'omuni* behave like carp, not [like those which move] in schools. They are fast, and stay on the side where the river is cutting under the bank, so they can go under it and hide. They sure are *slippery* ..."

As late as 1949, these suckers were being caught in commercial quantities in the Salt—up to six tons in a single spawning season. Even then, the humpback was declining because of dams that blocked their movements and that created temperatures below reservoirs too cold for spawning. During the last few decades, elderly Pima farmers express surprise if they even spot a single sucker while draining their ditches. Exotic fish, introduced into the reservoirs on the Gila and Salt, prey upon them. Biologist W. L. Minckley has shown that the humpback sucker was becoming scarce in the Salt and Gila rivers by the 1950s, and in the Colorado by the 1960s: "As with the Colorado squawfish, this unique animal is now extirpated in the Gila River basin, where it formerly occupied all large streams." Only recent reintroduction efforts have kept it from permanently vanishing from the Colorado, Gila, and Salt for good.

The only one I have seen alive was in an outdoor museum, where it hovered in front of an underwater window, staring back at the tourists who stared at it. Its mouth slowly opened and closed while it vacuumed the pond floor, and its grayish dorsal fin swept back and forth atop a bony, olive-colored hump. This one was younger than the few wild-born humpbacks left in the rivers. As Minckley has quipped, the viability of the humpback sucker population remaining in natural habitats is comparable to that in the nearby retirement community of Sun City, Arizona: "most are too old to reproduce."

Why, I wondered, did a fish become imperiled after it had survived harvesting by streamside desert dwellers for centuries? Minckley has identified the humpback and five other large fish from the prehistoric Hohokam village of Snaketown on the Gila. Although never abundant in Hohokam sites,

Minckley believes they were easily accessible to irrigation farmers in both prehistoric and early historic times: "Diversion of water and drying of a canal segment made them simple prey to man and other animals."

When the Spaniards first contacted the River Pima culture in the 1690s, they described this tribe as subsisting on "the innumerable fish that abound in the river," a diet *supplemented* with maize and beans. Is it a coincidence that the many varieties of maize and beans formerly grown by the Pima were lost from their diet at about the time that the fish began to vanish?

I think not. Over the last decade, I have sporadically assisted my former teacher, Amadeo Rea, with his life work of documenting the changes in the economy and habitat of the River Pima. In the process, I have become convinced that both cultivated and wild food resources were imperiled whenever the cultural landscape radically changed along the Gila and Salt rivers. These changes initiated the abandonment of some traditional Pima crops long before improved, modern crop varieties could have been introduced to supersede them. Like the Hohokam before them, the Pima lost everything whenever they lost control of the river that fed their canals. When the river was kept from recharging floodplain habitats with moisture and nutrients, both wild and domesticated resources would decrease in frequency and in productivity. Wildlife would also become scarce. The drying of a river and the ditches it fed would trigger the drying-up of gene pools.

Compared to the soil erosion evident on stretches of the downcut channel of the Gila River, such genetic erosion is relatively imperceptible. Few of us have the foresight to catch the fall of genes from a single crop, let alone recognize the impending collapse of whole farming communities as it begins to occur. If we do, we are initially hard-pressed to associate these losses with their ultimate causes.

Today, it is hard to imagine how either the Hohokam or Pima ways of farming worked. Only six of the 1750 miles of Hohokam canals in metropolitan Phoenix remain intact, the rest having been eroded, bulldozed, and built upon over the last century. After digging and sampling two miles of backhoed trenches across Hohokam canals, archaeologist Bruce Masse recovered only three pollen grains of a single domesticated plant: maize.

Inspection of a grain of pollen through the ocular lens of a microscope hardly offers a profound sense of the former agricultural grandeur of a region. Over much of the continent, where fields once flashed green with a dozen native crops, palynologists find a few pollen grains of domesticated plants per gallon of soil. Where prehistoric women prepared succotash day after day, archaeologists are lucky to discover more than a handful of dessicated corncobs and squash rinds. Where vibrant farming cultures nurtured fields for generations, the soil is now covered by offices, feedlots, slaughterhouses, subdivisions, and shopping centers.

To recover a sense of what native farming once was, we must cultivate our imaginations—allow pollen grains to bloom, becoming mixed fields. The asphalt, metal, blood, and manure of the last few centuries must be stripped away, exposing fertile soils once again. Seven thousand years of plant culture, buried beneath our feet, must be unearthed. America before Columbus was not a wasteland, nor an untouched wilderness. It held home ground for farmers; vast territory for hunter-gatherers; and places where farming and foraging fused themselves into the same cultures. To feel at home here, to learn from our predecessors on this continent, each of us must kneel on the ground, put an ear to the earth, and listen.

II.

In the 1530s, the lost party of Alvar Núñez Cabeza de Vaca arrived at an Indian village on a large river somewhere in Texas; there, they were greeted by an odd sound:

> At sunset, we reached a hundred Indian huts and, as we approached, the people came out to receive us, shouting frightfully, and slapping their thighs. They carried perforated gourds. . . which are ceremonial objects of great importance. They use them only at dances, or as medicine, to cure, and nobody dares touch them but themselves. They claim that those gourds have healing virtues, and that they come from Heaven, not being found in that country; nor do they know where they came from, except that the rivers carry them down when they rise and overflow the land . . .

Gourds such as those may have been the first plants cultivated north of Mexico. Around seven thousand years ago, they apparently were grown in small gardens in southern Illinois, as the archaeological reconstructions of David and Nancy Asch have demonstrated. And yet, this crop, in and of itself, hardly changed the diet of Eastern Woodland hunter-gatherers. They continued to rely on hickory nuts, walnuts, and groundnut tubers as their staples for many more centuries. Nonetheless, gourd cultivation may have begun the process of seed selection, saving, and sowing that has involved more than two hundred and thirty generations of Native Americans north of Mexico. The gourds that Cabeza de Vaca saw were recognized as a gift— something peculiar in the plant kingdom, and something fitting human needs.

We do not know what species of gourd the Indians danced with that day, four hundred and fifty years ago. But archaeologists have recovered from ancient Woodland sites in Illinois both the bottlegourd, *Lagenaria*, and *Cucurbita pepo*, the same species as the wildly colored ornamental gourds and jack-o'-lantern pumpkins found in fruit markets at Halloween time. In early prehistoric times, *pepo* fruits were undoubtedly smaller, with harder rinds. These gourds could have provided bowls and canteens to people who lacked pottery, or supplied greasy seeds to mix with other vegetable foods.

The early gourds grown along the Illinois River may not have been native that far north in the Mississippi watershed. Although egg-shaped pepos are sometimes found today on riverbanks from Illinois through Missouri to Texas, they are usually thought to be escapees from recent cultivation. Some geneticists and archaeologists speculate that gourds originally were traded northward, possibly from Tamaulipas (just south of Texas), where truly wild relatives of these gourds persist to this day. Other scholars, such as Charles Heiser, leave open the option that gourds may have been independently domesticated in the Eastern Woodlands and in central Mexico.

Regardless of their place of origin, cultivated gourds and related squashes have undergone considerable evolution in North America. Over seven millennia of growing in what is now the United States, *Cucurbita pepo* did indeed become native to the northern soil. Thanks to Native American horticulturists, squashes and gourds floresced into numerous folk varieties not found anywhere else in the world.

It was once thought that the entire vegetable repertoire of early North American gardeners was imported, part and parcel, from Mesoamerica. Although it still serves certain geopoliticians to claim that North America borrowed all its crops from other lands, this notion finally has bitten the dust. Much earlier than the arrival of Mexican corn varieties about 1500 years ago, Eastern Woodland horticulturalists domesticated several seed plants on their own, such as sumpweed, giant ragweed, and sunflower. Combine these natives with later introductions of common beans, bottlegourds, tobacco, and goosefoot, and Eastern Woodland farmers had a wide array of crops in use by 1000 A.D.

One of the domesticates native to North America is the sumpweed or marshelder. Known to scientists as *Iva annua*, this sunflower analog was found in association with humans in the Mississippi watershed nine thousand years ago. By 4000 B.C., this oily seeded annual was intentionally cultivated and in widespread use throughout the lower Midwest. And, with cultural selection over the following two thousand years, its seeds show impressive size increases. Knotweed, maygrass, little barley, giant ragweed, Jerusalem artichokes, and sunflower may have become common garden plants in the Eastern deciduous forest zones, although the latter was the only food crop that emerged as a domesticate fully distinct from its wild ancestor.

Some scholars guess that the success of the domesticated sunflower may be what led to the demise of the sumpweed crop. In fact, sumpweed cultivation not only declined in late prehistoric times; the Indian-selected variety ultimately became extinct. Around the time of the first European contact, large-seeded sumpweeds drop out of sight. In 1972, when archaeobotanist Richard Yarnell described *Iva annua* variety *macrocarpa* as "an extinct American cultigen," he added these speculative remarks:

"It is possible that sumpweed cultivation developed earlier and had the initial advantage of being well adapted to the wetter climate of the Eastern Woodlands. It may be that sunflowers ultimately outdistanced sumpweed in productivity and totally displaced it as an oil source toward the middle of the second millennium A.D. Both plants apparently were displaced as major food sources by corn, perhaps by the latter half of the first millennium A.D."

Sunflower achenes (seeds enclosed in shells) were about twice the size of sumpweed achenes, and so sunflowers may have been favored over sumpweeds for that reason. But why did the two crops continue to coexist in the same fields for thousands of years before one fell by the wayside? And while the sunflower was displaced later by corn as a major staple, why did it not disappear as a cultigen as well?

As ethnobotanist Richard Ford has observed, sumpweed actually continued to be grown "in the central Midwest until historic contact (with Europeans)." Yarnell alludes to its presence in Ozark caves as late as 1430, and does not rule out a terminal date of 1670 A.D. Was sumpweed made obsolete by the sunflower in prehistoric times, or did it rapidly vanish when Europeans began to vanquish the East? Wasn't native farming altogether disrupted elsewhere by European-introduced diseases, pests, weeds, and competing crops? If sumpweed persisted as a minor crop until after the European arrival on this continent, how would it have fared when local farmers were being decimated by new illnesses, and the landscape was being changed in innumerable ways?

I don't know for certain why sumpweed achenes aren't eaten anymore, for the details of its demise are lacking. But one thing is clear. Today, the pool of genes left from domesticated sumpweeds is as dry as a Hohokam canal.

III.

What can't be learned from the sumpweed story in America's heartland is amply evident in the desert where the Hohokam canals once brought water to sixteen species of desert crops. Here, it is possible to relate the waning of many prehistorically cultivated plants to particular periods of environmental stress or social upheaval, and sometimes to ultimate causes.

Along the Gila and Salt floodplains of central Arizona, twenty-five hundred years of crop history have been tracked by some of the finest ethnobotanists who have worked in the New World: Edward Palmer, Edward Castetter, Hugh Cutler, Vorsila Bohrer, Jonathan Sauer, Amadeo Rea, Charles Miksicek, Suzanne Fish, Robert Gasser, Adrianne Rankin, and Jannifer Gish. A remarkably complete sequence of crop introductions and extinctions can be pulled together from these scholars' identifications of excavated plant macrofossils, pollen, interpretations of historic documents, and collections of oral histories. They record radical changes in the biotic diversity of a land described in 1795 as "good for all kinds of seeds and plantings." Today, the floodplains of the middle Gila and the Salt harbor a paucity of crops, and far fewer wild plants and animals than ever before.

To reconstruct where the Hohokam crops came from and where they went, we must travel back before canals of any size were constructed along the Gila and Salt rivers: to a period prior to the florescence of the Hohokam.

Maize became a significant foodstuff in southern Arizona in Late Archaic times, between 1000 and 700 B.C., during a period when floodplain soils were naturally accumulating rather than downcutting into gully-like arroyos. The earliest corns in the Southwest, a flinty chapalote and reventador popcorn, are no longer found in the United States, although I've collected both from Indian farmers just a few hundred miles south into Mexico. But the other race of maize that arrived in the Gila and Salt valleys early on—the floury eight-rowed sixty-day corn—persists until this day. These corns may have initiated crop production adjacent to the Gila and Salt rivers, where shallow ditches diverted runoff or streamflows into small fields cleared on the lower floodplains.

Between 300 B.C. and the time of Christ, a small group of native horticulturists began to utilize a set of technologies previously unknown in the Gila-Salt region. Labelled the Hohokam by archaeologists this century, this prehistoric culture began producing pottery, textiles, stone and ceramic sculpture, and houses in a distinctive, sophisticated manner. Their canals of water diverted from springs, marshes, and rivers allowed for a greater intensity of crop production than what rainfed or runoff farming might provide in a desert region. Through the end of the Pioneer Period between 600 and 700 A.D., they gradually added several minor crops to their cornfields: common beans, cotton, agaves, and little barley. At the same time, they continued to rely upon a wide variety of wild foods, such as mesquite, cacti, and winter annuals. Some of these wild foods, such as greens, became more intensively harvested from ditchbanks and fields.

During the next two periods of Hohokam culture, the Colonial and the Sedentary, these "backwater farmers" spread into a wide variety of environments across fourteen thousand square miles of southern Arizona. Archaeologist Dave Doyel observes that from 600 to 1100 A.D., "The Hohokam underwent a time of population growth, regional expansion, and developing complexity in their material products. . . . Settlements in the central region, such as Snaketown, also continued to grow, approaching [240 acres] in area and populations of approximately one thousand people. . . . Large canal systems serving multiple villages became common. . . . Some of these ancient ditches, excavated by the Hohokam from hard desert soils with nothing more than pointed wood sticks, measure over five [yards] wide and three [yards] deep."

Finding their way into the Hohokam larder were new crops from the south such as tepary beans, limas, tobaccos, and squashes. Around 1000 A.D., when a warmer, wetter period began, ethnobiologist Charles Miksicek believes that climatic conditions became more favorable for Mesoamerican domesticated crops. Two grain amaranth species became prominent, and tropical jack beans made a brief appearance in the region. There is tantalizing evidence that a Mesoamerican goosefoot may have also been introduced, here and in the Ozarks and Great Plains. The little barley plant's grains show signs of incipient domestication after 1000 A.D. The spectrum of crops available to the Hohokam had radiated into a rainbow of seed colors. A thousand years ago, the Hohokam had the most diverse crop complex of any culture living in what is now the United States.

But this cultural flowering, and the moisture that allowed its growth, was of short duration. Around 1150 A.D., a drought may have starved the fields fed by river-diversion canals, and there are increasing signs of the build-up of salinity in Hohokam fields. By this time, some of the canals had been consolidated into more extensive irrigation systems. Because so vast a canal network had been built over an eighteen mile stretch of the Salt River, farmers could easily deplete the river of its entire water supply during months of low flow. A breakup and period of social disequilibrium among the Salt River Hohokam between 1050 and 1200 A.D. may have been aggravated by this water crisis. About that time, Snaketown was abandoned; pottery and other cultural remains dating from the Late Sedentary period are altogether lacking from that large settlement on the Gila.

As more people migrated to the well-watered reaches of the Salt from surrounding drought-stressed areas, huge new canals were constructed that connected several older irrigation areas. Still, water conflicts and times of scarcity must have arisen. Mesoamerican crops such as amaranths and jack beans dropped out of the prehistoric archaeological record of the region.

"But why?" I recently asked Charlie Miksicek, as we went over mounds of his carefully-gathered data on the kitchen table of his Tucson home. He paused for a moment, and then answered: "I think climate has a lot to do with it, but so does cultural upheaval. After 1150, the people there don't behave like the earlier Hohokam. They were retreating from some of their older villages, leaving things behind . . . "

Archaeologist Dave Doyel's analysis echoes the same feeling. It was not a time when people simply dropped a few moisture-loving crops, reorganized their canal systems, and went on again without mishap. The greater population concentrations clustered on the Salt and the Gila led to what some call "subsistence stress," and what others refer to as "unregulated competition for water." Downstream villages were disadvantaged. Bruce Masse sees "the dissolution of a regional system of water management and . . . resultant abuses of the water supply upstream."

But the movement toward longer, larger canals in communities concentrated upstream may have led to greater vulnerability. Dendrochronologist Don Graybill has integrated tree ring analysis of climate change with archaeological interpretations of the Salt River environment in the fourteenth century. His reconstructions of rainfall and river water availability for the Salt River watershed rough out a disturbing picture: "From 1300 to 1350, a period of decreasing streamflow occurred that was nothing like what the sedentary Hohokam had experienced before," Graybill explained to me. The watershed dried out, and perhaps lost a considerable density of its plant cover. "Then," Don went on to say, "there were several extremely wet years: 1354, 1356, and 1358. A high spike in stream flow. A wet period after a long dry one. The conditions through the mid-1350s were the kind that could have resulted in very major damage to canal intakes down in the valley."

If the land was barren and brittle enough from the long drought, such wet years could have carried monstrous floods off the uplands in the watershed above the Hohokam farming villages. These floods may have cut the river channel below the level of the canal intakes, or washed them away altogether. By the time another devastating flood occurred in 1382, much of the irrigation systems on the Salt had likely been abandoned, and human populations had begun to relocate themselves. Although many Hohokam archaeologists still maintain that the Hohokam disappearance occurred one or two centuries later, few disagree that the growth of this civilization had been curtailed by the 1350s. Dave Doyel feels that these natural disasters were among the forces that ultimately "stressed the Hohokam system."

Already overspecialized and hyperdependent on their huge canals, the Hohokam began abandoning their riverine farming villages, canal systems, and ceremonial centers. Between 1350 and 1400, the Salt and Gila valleys were depopulated. The agaves and native barley patches that had been plentiful were gone by the time the first European observers set foot in this region. Eleven of the sixteen prehistoric crops somehow hung on, persisting into historic times, but they too may have suffered from varietal erosion during the Hohokam collapse.

Hohokam. It is derived from the Piman term *Huhugam O'odham,* "the people who have vanished," or "exhausted people." In Pima legend, they were on this earth before the Pima emerged. They began to disappear into a hole in the ground when Coyote said something to them that kept them from vanishing altogether. They are known to us today by their rock art, their large ruins, and their widely-scattered pottery shards. We know which of their crops were lost, and how pottery designs rich in images of aquatic birds were never to be made again. What we'll never know is how the last Hohokam farmers felt when they abandoned their fields along the canals built by their forebears.

IV.

"The green of those Pima fields spread along the river for many miles in the old days," recalled the River Pima leader, George Webb, from his home in Gila Crossing. But that was in the time "when there was plenty of water. Now the river is an empty bed full of sand. . . . Where everything used to be green, there were acres of dust, miles of dust, and the Pima Indians were suddenly desperately poor."

In those words, George Webb telescoped the history of the last three hundred years in his homeland, the former home of the Hohokam. During those three centuries, the Pima have lost at least seven crop species that were introduced to the Southwest prehistorically. Seven native varieties of five New World species survive precariously today, in the dooryard gardens and small fields of the River Pima: sixty-day corn, white and brown teparies, mottled limas, narrow-seeded bottlegourds, and striped cushaw squash.

It was not only the native crops of "subsistence farmers" that died of thirst. The Pima's once-thriving export economy based on Spanish-introduced wheat faltered also. In *Once a River*, Amadeo Rea has recorded twenty-nine species of birds that were extirpated in Pima country over the last century, and he counts another dozen bird species that dramatically declined because of habitat deterioration there. Fish, like the humpbacks, and river-loving mammals were left high and dry. And as Rea has since discovered, the number of wild plant resources lost from the Gila floodplain was of an even greater magnitude.

Without these resources, the life of the *Akimel O'odham*, "River People," changed radically over the last one hundred twenty years. Yet as Henry Dobyns has suggested in his ethnohistory essay, "Who Killed the Gila?" the erosional processes that disrupted agricultural landscapes on the Gila River floodplain were set in motion by the Spanish much earlier. The Spanish introduction of European diseases and livestock triggered changes in the intensity of management of the Gila floodplain. Along one stretch of the Gila, Rea has recorded that the number of River Pima *rancheria* settlements fell from thirty to just three in the first century after Spanish contact, presumably because of the depopulation caused by smallpox and other pandemics. Unable to manage the Gila floodplain with the intensity of labor that they had previously, the River Pima perhaps more readily converted croplands that had been abandoned to pastures for newly-arrived livestock. By the mid-nineteenth century, cattle populations in the Gila watershed had overgrazed the vegetation to the degree that hydrological conditions were severely altered.

Well in advance of the Civil War, overgrazing as well as woodcutting and beaver-trapping in the Upper Gila began to change gentler streamflows to flashfloods capable of eroding out the floodplain fields of the Pima. Yet there was a time lag before these disturbances actually began to disrupt the agricultural environments tended by the Pima. During drought and flood sequences much like those that wreaked havoc during Hohokam times, the repercussions of many environmental disturbances became concentrated on the middle Gila floodplain, where most Pima farms were located. Arroyo cutting and aquifer draining were initiated as early as mid-century, but accelerated by the 1870s.

Then, in 1867, an additional sort of pestilence fell upon the River Pima of the middle Gila and the adjacent Salt. That year, Jack Swilling, an entrepreneurial alcoholic and morphine addict with vision enough to notice the old Hohokam canals, organized his Anglo neighbors to divert water out of the Salt River above the Pima villages. He began to "reclaim the desert," an effort continued today by the massive Salt River Project.

Upstream from most of the Pima villages on the Gila, other recent arrivals to the desert opened up large canals during the drought in the early summer months of 1870. These Anglos poured onto their fields an excess of irrigation without returning their tailwater to the river for others to use. By 1873, three hundred Pima and Maricopa Indians left the Gila for good, hoping that in a new settlement upstream from Phoenix they would avoid such conflicts. The situation did indeed become far worse on the Gila, because settlers in Florence began to expand their new farming colony upstream. Ethnohistorian Edward Spicer wrote that "By 1887 the irrigation canal constructed to take water out of the Gila River utilized the whole flow. No water reached any of the Pima fields downstream."

Only where the Pima could rely on the tributary flows from the Salt or the Santa Cruz was any native farming still possible. Fields that once held a mix of crops were left too dry to support anything but the hardiest of weeds. What happened, then, to the seedstocks usually saved in Pima storage baskets and pots? Did they remain in storage, unplanted, until they lost their viability? Were they sold off to those who still had water? Were the grains and beans eaten up as food supplies dwindled?

V.

In the 1750s, when missionary Juan Nentvig visited the depopulated Pima, he claimed that "so much cotton is raised and so wanting in covetousness is the husbandman that after the crop is gathered in, more remains in the fields than is to be had for a harvest here in Sonora." By 1873, this fiber crop was already declining. When botanists finally collected a few samples just after the turn of the century, they claimed that it had all but died out completely. In 1901, Frank Russell claimed that there was so little cotton that there was not enough in Pima villages to finish making even one piece of cloth on a small loom. He added that "the Pimas no longer spin and weave; the art is dying with the passing of the older generation."

Although Egyptian cotton production began in central Arizona in 1908, it did not directly replace the fine-fibered native cotton. The aboriginal Pima cotton was nearly extinct when modern "Pima" cotton was selected from a single superior plant of the Egyptian species in 1910. The Pima cotton in our clothes today was not commercially produced until 1916. By that time, the only remaining authentic collection of aboriginal Pima cotton was restricted to plant breeders' collections in agricultural field stations.

The last remaining grain amaranth apparently perished between 1870 and 1890. Varieties of common beans, squashes, and a small grain called *kof*—perhaps a relic goosefoot surviving from prehistoric times—ceased to exist even in hand-watered kitchen gardens. Foods that had been mentioned in the Pima creation myth were never again grown, prepared, or eaten after 1900.

Following the turn of the century, the remaining wet places on the reservation dried up. Cultivated acreage decreased to less than a third of what it once was. From 1898 to 1904, droughts and upstream diversion of water by Anglos kept the Pima from producing any crops at all. Groundwater pumping was soon initiated. It quickly depleted remaining springs and seeps near farmlands. Mesquite and cottonwood forests died as the water table dropped below the reach of their roots. In their stead, the exotic salt cedar took hold.

Men such as George Webb were forced out of farming in the 1930s due to lack of irrigation water, even though the San Carlos Irrigation Project promised them a return to better days.

"When the dam was completed there would be plenty of water," Webb remembered. "And there was. For about five years. Then the water began to run short again. After another five years, it stopped altogether."

It was that time that Johnny Cash sang about in the ballad of the Pima

Indian hero at Iwo Jima, Ira Hayes: "Down the ditches of a thousand years, the waters grew their crops / Until the White Man stole their water rights, and the sparkling waters stopped / Then Ira's folks grew hungry, their land grew crops of weeds / When the war came he volunteered, forgot the White Man's greed."

The year when the Webb and the Hayes families had to abandon their fields, Edward Castetter and Willis Bell began fieldwork on Pima agriculture. They caught the tail end of garden cultivation of two native tobaccos, and most if not all of the native common beans. Mottled limas had been absent from Pima marketplaces for two decades, but were then reintroduced by the USDA station at Sacaton, the headquarters for the Gila River Indian Community. However, the Indians who had access to San Carlos Project water were required to grow alfalfa and barley rather than their traditional crops, as part of a government-promoted soil building program.

The amount of farmed land on the Gila River Reservation increased in the late 1930s, but this was due to the addition of the Agency farm controlled by the Bureau of Indian Affairs, using some twelve thousand acres of unallotted reservation lands. Take away this acreage, and Pima families were farming less land for themselves in the thirties than they were at the time of the Civil War. And, because of the earlier constraints put on them by a BIA-enforced allotment program, Indian farmers had less land to work with. Between 1914 and 1921, each member of the Pima tribe had been allotted just ten acres of arable land. Twenty years later, the progeny of those tribal members were having to divide up that land. Today, some of the original allotments must be split among a hundred or so descendants.

For those who had only a few acres at their disposal, and no credit or equipment with which to develop them, it was easier to turn to wage work. Many Pima families left their whittled-down farmsteads to pick cotton for Anglos in the Casa Grande Valley to the south of the reservation, never to return to family farming again.

VI.

Another, far more debilitating kind of allotment program had happened on most Indian reservations, one that the River Pima had successfully resisted.

This allotment program, initiated by the 1887 Dawes Act, is considered by many historians to have begun the decline of Indian farming in most parts of the United States.

Economic historian Leonard Carlson has documented the failure of Dawes Act allotments, which presumed to encourage Indians to initiate family-based farming enterprises, by means of a land development scheme superficially similar to the Homestead Act. In many areas, white settlers were also allowed to purchase unallotted Indian lands that had formerly been held in trust for the entire tribal community. In an argument that thinly disguised their motive, real estate developers asserted that "protecting Indian ownership of unused land would encourage idleness." Provided with economically industrious Anglo neighbors, the Indians would be more rapidly assimilated into the dominant society.

In one year alone, 1891, the Indian Commissioner Thomas Morgan sold off one seventh of all the Indian lands in the United States to white settlers, some 17,400,000 acres. Other kinds of ownership transfers were common during this period as well. All told, from the 1887 implementation of the Dawes Act, to the termination of allotments and sales of Indian territories in 1934, some eighty-four million acres of land passed out of Indian hands. This amounted to 60 percent of all the land that had been held in trust or through treaties on behalf of Native American peoples when the Dawes Act was passed.

Much of the arable land that Native Americans had formerly utilized for farming, hunting, and gathering was thus usurped by others. Where supplemental water was essential to avoid crop failure, Indians often found irrigation projects biased toward their non-Indian neighbors. Once their land and water were taken away, it is no wonder that Indian farming declined, and with it went much of the remaining native crop diversity.

Nevertheless, most Indian people remained close to the land, even though many of their families were suffering from both a declining cash income and a deteriorating resource base. In 1910, 95 percent of the Native Americans in the United States lived in rural areas. This percentage dropped only to 90 percent in 1930, and 78 percent in 1960. During the era of the Dawes Act, its proponents claimed that allotments would encourage more Indian farming. However, the percentage of the Indian population employed in agriculture declined from 74.7 percent in 1910 to 64.5 percent in 1930.

Perhaps some sociologists would claim that this decline was not catastrophic; after all, it simply mirrored the move of the general American population away from the farm. During the same period, the overall farm population of the United States slid from 34.9 percent to 24.9 percent of the total population.

Many Native American farmers, however, were irrevocably losing control of croplands that had been in their families since long before the arrival of Europeans. In 1910, about twenty-two thousand Indians worked their own farms while another 26,500 served as farm laborers. By 1950, despite considerable growth of Indian populations, the number of Indian farmers had dropped to 14,300, and their farm labor force to 14,100. By 1982, only forty-seven hundred Native Americans were full owners of farms, another seventeen hundred were part owners and seven hundred fifty were tenant-operators. In total, about seven thousand Native Americans are managing farms at present. Less than five hundred of these Indian farms nationwide grow any sizeable mixture of crops, including vegetables, staple grains, and beans for self-consumption. Of course, not even all of these farms focus on traditional crops native to their region.

These data from census statistics contrast sharply with the inflated numbers from USDA and BIA reports promoting "new initiatives" to help Indian farmers. For instance, the USDA today claims participation in its programs by fifty thousand Native American agriculturists—more Indian farmers than the census bureau has counted since the turn of the century! A BIA work group on Native American agriculture recently boasted that thirty-three thousand individuals and tribal enterprises are involved in farming or ranching on Indian reservations, utilizing nearly fifty million acres of land. Those figures seem impressive until it is realized how little of it must be anything more than introduced forage and livestock production.

How much of that "Native American farming" effort draws upon crop resources traditionally utilized by the cultures involved? Neither the USDA nor the BIA bureaucrats track the actual use of native crops. To be sure, these agencies have hardly ever promoted those resources either.

In fact, several years ago, USDA policy makers published a statement of their legacy entitled, "The Lack of Native Crops in the United States." In 1979, and again in 1984, the official USDA policy statement on plant genetic resource conservation began with this preposterous assertion:

"If American consumers were asked to live on food from crops native to the United States, they would probably be shocked that their diet was limited to sunflower seeds, cranberries, blueberries, pecans, and not much else. . . . Tobacco would be available, but they would have no cotton. . . . [The] resources that support our domestic food and fiber production [today] are imported."

In a meeting with USDA officials, Kent Whealy of the Seed Savers Exchange pointed out why so many of our plant genetic resources are indeed imported from developing countries: "There has never been a systematic, large-scale search for [crop] plants *within* the United States; the USDA has always done its explorations outside the U.S."

Further, USDA policy claims that *in situ* conservation of crops—which could be done by providing incentives to native farmers to keep growing their remaining traditional varieties—is too unstable, risky, and ineffective, and therefore beyond the scope of its plant preservation concerns. The only real policy that our Department of Agriculture has for the native crop legacy of our country is "a policy of neglect."

VII.

Fortunately, tribal governments no longer believe fatalistically that this must be their policy as well. Among the Iroquois, the Sioux, the Mississippi band of the Anishanabey, San Juan Pueblo, the Winnebago, the Tohono O'odham, the Navajo, and other tribes, there have emerged community or tribal projects to conserve and revive native crops as cottage industries for their rural-based tribal members. On both reservations where the Pima live, they have initiated tribally-supported farm efforts to increase the supplies of traditional crop plants.

Perhaps the most appealing of these efforts has been the Agricultural Resources Project of the Salt River Indian Community. A few years ago, a survey of tribal members indicated that it was hard to obtain the native foods they favored. And of more than sixty families contacted, only four said that they were currently growing native crops.

The community obtained foundation assistance to design a project that would provide nutritious foods as well as new sources of income to its Pima and Maricopa members. Three young Native Americans came on as the project staff: Darren Washington, Angie Silversmith, and Berkley Chough. On five acres of land, they experimented with six commercial vegetable varieties, as well as twenty-five traditional desert crops provided to them by Native Seeds/SEARCH.

As NS/S contact Kevin Dahl recalled from his conversations with Angie and Berkley, "When bugs became a problem, they overwhelmed the commercial vegetables while hardly touching the native varieties. Not all the native crops were a success, however. . . . [But] it was the Gila Pima corn, Pima cushaw squash, Papago sugar cane, and Papago dipper gourds that flourished. [And] Tepary beans proved to be an outstanding success."

These crops have been grown twice and sold to community members, who expressed great interest in them. On a reservation where more than three quarters of the farmable land has been rented out to Anglo lessees for decades, the Pima and Maricopa have newly demonstrated the value of crops from their own tribal legacies. However, the future of the Agricultural Resources Project remains uncertain; foundation assistance has terminated, and it is up to the tribal government and community to reinitiate plantings without outside support.

From the renewed interest generated by this project, the Salt River Indian Community helped Scottsdale Community College sponsor a conference called "Native American Agriculture—A Critical Resource" in October, 1987. In its position statement, the conference coordinators reminded participants that even today, "Indian land and its potential agricultural use are in danger of being lost. The most important economic resource available to the American Indian is the land and its agricultural potential. Properly using agricultural land within Indian reservations is one of the greatest challenges now confronting Indian people and tribes."

Out of the conference came a proposal for an American Indian Agricultural Resource Center. To be located less than fifteen miles from the old Hohokam canals on the Salt River, this center would search for ways to ensure that tribal farmlands will be "wisely utilized and preserved for future generations." In light of what has already been lost, this proposal seems a century late. Yet the little seed and land that remain are valuable enough that the Pima wish to guarantee their survival. The feeling that something still precious remains also pervades the Park of the Four Waters, where remnants of some eighteen prehistoric irrigation canals run parallel to a modern one heading off toward downtown Phoenix. No matter that less than a hundredth of the original reach of Hohokam canals remains intact. No matter that dirt bikers use the eroding crest of the canal bank as a jump course. No matter that jets fly overhead every few minutes, obliterating any silent respect for the past that one tries to muster. Those ditches are monuments.

It does not require much quiet contemplation to be awestruck by what went before us. The Hohokam canals humble us as do the Mayan pyramids, the Sistine Chapel, or the Great Wall of China. If those who are involved in historic preservation wish to celebrate a great engineering feat from the pre-Columbian agricultural heritage of North America, let them provide greater protection for these canals.

But let us remember that centuries ago, this irrigation system provided grain and beans and fish for the bellies of people. It would be a hollow kind of historic preservation if sixty-day flour corn, tepary beans, and humpback suckers became extinct while the earthen walls of the ancient ditch were preserved as public monuments. Let us not overlook the monumental contributions of the crops far more ancient than the European discovery of this continent, foods that still have the power to nourish us. We must keep them alive.

CHAPTER FIVE

A Spirit
Earthly Enough:

Locally Adapted Crops
and Persistent Cultures

I.

They were planted on the edge of the mesa. They were sandstone houses, and I wove down the sandy roads and trails that snaked between them. From their roofs, juniper smoke rose, permeating the air. I could see both fuelwood from the mesa top and coal from its siltstone innards in piles beside the front doors of these houses. The oldest houses were made of rough-cut slabs taken from the mesa itself. These slabs were laid down in an overlapping fashion, chinked with pebbles and mortared with a gritty mud matrix. Newer houses, perhaps crafted early this century, were built from rectangular blocks neatly cut from the same sandstone.

Some of the houses were linked into what is called a pueblo. Pueblo also means people, or better perhaps, community. Residents of this pueblo are

Hopi. The community school, where friends and I were about to teach, was founded by Hopi families. It was a school grounded within a community fixed in place: a mesa in sight of *Nuva Tuqui Ove*, the sacred San Francisco Peaks. The local children, it was hoped, would learn not just of facts and theories, but how to value certain materials or matters within their fitting contexts.

I scanned these sandy paths before settling into the classroom. I noticed that some young adults were leaving the community to work for wages, but that a few were going with older women to tend recently-planted vegetables in the spring-fed terraces below the mesa. It might seem odd that three of us *pahaana* outsiders had been invited to teach their children about native plants, for there was a wealth of folk botanical knowledge still residing in this community. But we had not been asked to teach about plants as the elders would do. We had been encouraged to link what was already familiar to that formidable phantom named *Science*. To build this bridge, we simply started with those plants closest to home, Hopi crops, and we spoke of their adaptations to sand, wind, and drought.

We took our teaching tools into the classroom to set up before the students arrived: sand, seeds, posters, and other paraphernalia for experiments. The posters were oversized diagrams of corn and bean seedlings, naming their different parts. I had also built an upright box, one side of it a window, to fill with local dune sand. It looked like the "ant houses" in which, as a schoolboy, I had watched ants wander down tunnels in miniature work parties. Instead of stocking this box with ants, however, we had planted half of it with sand-adapted Hopi blue flour corn, and half with a commercially-available Bantam sweet corn.

A Hopi teacher walked by while we were setting up the posters on the blackboard.

"What are you planning to do with that drawing of bean sprouts?" she inquired.

"Teach the kids about seedling shapes and parts," I replied.

"How about the box full of sand?"

"We'll measure the growth of corn roots and shoots as they expand in the sand," my friend Mahina explained.

"Hmmmn." The Hopi woman paused, brow furrowed. "The box, and growing plants in the sand, seem just fine. But the bean sprouts on the poster. . . . Unh uh."

"What do you mean?" I asked, confused.

The Hopi teacher sat down, gazing at the poster. "See, the children here are presented with bean sprouts, but at a special time. The katsina dancers hand them bundles of sprouts early one morning, as part of the nine-day Bean Dance sequence. The children take these gifts back to their families, and a traditional dish is fixed from them. That's when Hopi kids are supposed to see sprouts out in the open. I don't know if it can be any time at all . . ."

"Well, it doesn't have to be . . ."

"It's because they're part of something sacred. We have our own way of teaching the children about them. Would there be any way that you could teach the students what you want to without showing them the bean sprout poster?"

"Sure . . . I suppose so," Mahina replied. "What do you think, Gary?"

"If we can show the corn diagram . . ." I stammered, "then we can go and relate seeds to seedlings to whole plants. We can still talk about the differences between corn and beans, which you can see even in the seed stage. Then we can show the corn seedlings growing in the sand. . . . If that's okay . . . ?"

It was, judging from the mass of students and teachers clustered around the corn-filled sand box an hour later. That spring week, we looked at the seeds of Hopi corn and bean varieties. We processed them and ate them. We "play acted" their adaptations to sandy, windy lands. Each class graphed the growth rates of the two kinds of corn, comparing them. Mahina and my wife, Karen, accompanied by older Hopi women, also took the children down into the terraces to plant rarer native crops.

Botany had come home, rooting itself in locally observable phenomena. The children had their hands and minds working with crops that some of them might cultivate, cook, and consume for much of their lives. And, as three outsiders, we had learned how to place science in a particular cultural context, a lesson more valuable than the display of one bean diagram.

II.

In the parable of the sower, Jesus of Nazareth described the many seeds lost by failing to reach suitable soil, ending up stranded on the rocks, among thorns, or in the bellies of birds. Other seeds fell into good earth, where their growth was favored. For a fertile, productive life, there must be some kind of match between the adaptations of the seedstock, and the particular conditions of the soil where it is sown.

In *The Gift of Good Land*, Wendell Berry articulated this challenge at another level, addressing the problems of agriculture, but acknowledging a universal dilemma at the same time: "The most necessary thing in agriculture . . . is not to invent new technologies or methods, not to achieve 'breakthroughs,' but to determine what tools and methods are appropriate to specific people, places, and needs, and to apply them correctly. Application (which the heroic approach ignores) is the crux, because no two farms or farmers are alike; no two fields are alike. Just the changing shape or topography of the land makes for differences of the most formidable kind. Abstractions never cross these boundaries without either ceasing to be abstractions or doing damage. . . . The bigger and more expensive, the more heroic they are, the harder they are to apply considerately and conservingly."

Berry then contrasts the heroic ideals attributed to certain religious traditions with the daily practice of the tenets of those traditions. Such daily application means paying attention to particulars, letting a spirit of "connectedness" be manifested in one's work. In the Judeo-Christian tradition, Berry believes that there are biblical mandates for humans to exercise constraint in their use of the land and its verdure, but that few Christian communities consistently heed these mandates. Elements of a conservation ethic may be found in many religious traditions, but few who nominally embrace such beliefs are willing to constrain their own behavior. There may be little evidence of conservation if we simply look at their consumption of local resources and their pattern of land use.

Where do we find the communities that embody a daily practice of caring for the soil, its plant cover, and the creatures so often described as "God's gifts"? Is such an earthward practice evident in the lofty hierarchies of institutionalized religions?

More often than not, the practitioners who express their care for the health of the land are found on peripheries of spiritual traditions, in rural villages remote from the Vaticans, Jerusalems, Meccas, and evangelical headquarters of the world. Berry argues that the Judeo-Christian traditions, for the most part, have become too abstract, not *earthly* enough in the agricultural applications of their worldview. Yet this problem is not limited to Christianity; such otherworldly tendencies have propelled many of the "civilized" religious empires throughout world history.

The impression is inescapable that much of the destruction of lands and their plant diversity wrought by both Western and Eastern ecclesiastical fervor has been dismissed by the priestly class as purely insignificant. Environmental historians, however, have unearthed ample evidence of the failure of certain religious traditions in taking heed of the ecological constraints of the settings in which their civilizations emerged. In recent decades, much of the desertification of the Middle East has been blamed on Hebrew, Christian, or Islamic mismanagement of land resources, but perhaps its impetus predates, and is not limited to, these traditions alone.

Consider the Sumerians, who created what environmental historian Alfred Crosby calls "humanity's first real civilization" in the Mesopotamian cradle of religions. Crosby reminds us of the power gathered by Sumerians about five thousand years ago "in the flat lands around the lower reaches of the Tigris and Euphrates . . . in their crops of barley, peas, and lentils, and in their herds of cattle, sheep, pigs, and goats." And yet, their power also had its origin in the organic detritus which had periodically renewed the productivity of the "Fertile Crescent."

Tragically, this gift was wasted. The Fertile Crescent became ravaged by the soil erosion, salinization, and siltation aggravated by agricultural mismanagement. As annual seed plants and livestock herds were extended beyond their appropriate scale, the side effects of poor irrigation and overgrazing debilitated the Tigris-Euphrates watersheds. At Tell-Asmar, once fertile lands farmed thousands of years ago now sit below ten to eleven yards of silt.

I doubt that such a fact is ever discussed in courses on Western Civilization and Religion as they are taught today. Desertification is not recognized as evidence of negative feedback from poor practices associated with a certain belief system. In the New World, there are similar examples of sizeable agriculture failures among sophisticated civilizations. Watershed denudation combined with economic overspecialization may have been among the factors leading to the collapse of the Anasazi at Chaco Canyon and the Mimbres culture in southern New Mexico. Nevertheless, scholars can seldom pinpoint the driving forces of an agricultural collapse, and therefore remain justifiably timid when it comes to relating such catastrophes to the ecological sensitivity of cultural belief systems. Others simply ignore the notion that the overexploitation of land and life could be the direct effect of a culture's value system; the study of one is rarely connected with the other.

However, some currently existing cultures contend that the way one farms and cares for local resources has *everything to do* with the spiritual life of the community. Certain Native American villages are exemplary of this worldview, for their clans have tended the same fields and gardens for centuries without exhausting their fertility or plant diversity.

These cultural communities offer us insights into the mutually reinforcing connections between a sufficiently earthly spiritual life, and skilled concern for the ecological integrity of food-producing land and plants. Yet we should not idealize them as living in a static balance with nature, nor assume that all individual farmers within them have the same exact skills or the same allegiance to community values. These communities are dynamic and internally heterogeneous, but a shared set of traditional values and practices have nevertheless persisted in them over the years. Although not unchanging, their relative resilience has resulted in enough environmental stability to keep a diversity of locally adapted crop varieties alive. As we shall see, the means by which these cultures maintain crop diversity are many, as are the benefits ultimately derived from this diversity.

III.

Locally adapted cultivated plants are variously referred to as folk varieties, land races, heirloom vegetables, crop ecotypes, or *razas criollas*. They represent distinctive plant populations, adapted over centuries to specific microclimates and soils. They have been selected also to fit certain ethnic agricultural conditions; the field designs, densities, and crop mixes in which they have been consistently grown. Aesthetic selection has also taken place, as the taste, color, and culinary preferences of a particular culture have favored the forms and chemical characters of some plants over others.

The emergence of crop ecotypes occurs most frequently in the isolation of remote agricultural areas. Where intercropping is practiced over decades, a co-adapted complex of crops evolves, each with genetic traits that reinforce those in its companion crops. Each crop may also develop multigenic adaptations to the prevailing stresses in the region, particularly if it is given enough exposure to such stresses over a long period of time. Since heirloom vegetables are by definition those passed generation to generation through family or clan, they are best represented in cultural communities where a thread of continuity has woven through the centuries.

In short, one can hardly create a dune-adapted Hopi crop in a lab overnight. A biotechnologist can't simply transfer one or two genes from variety A to variety B to get the same adaptive qualities. I have no idea how many genes make Hopi blue flour corn distinct from visually similar seeds originating elsewhere, but its responses are markedly different when grown in the same plots with the others. Ethnobotanists Rita Shuster Buchanan and Robert Bye have demonstrated these differences by growing Hopi blue flours side by side with Tarahumara blue flints from Chihuahua. Plant architecture, cob placement on the stalk, days to maturity, ear weight, and yield were considerably different, even though these two races diverged from the same evolutionary line not so long ago. Their differences in taxonomic characters are unremarkable, but the two varieties each bring into play different morphological and physiological strategies for survival when sown in the same semi-arid environs.

Early in this century, USDA botanist G. N. Collins discovered that Hopi corn had morphological adaptations to deep-seeded, clumped plantings in sand dune environments. Among these adaptations was the robust seedling's rapid elongation between the root and the first foliage leaf of the developing plant, allowing it to emerge from beneath twelve inches or more of sand!

Hopi farmers did in fact plant their corn seed eight to twelve inches deep, for the sand stays moist at this depth. The seedlings that emerged in May could endure on the residual moisture from winter and spring storms, until the summer rains began. Collins suggested that Hopi and Navajo corns from sandy fields were unique in their ability to tolerate deep plantings, but he did not formally compare their responses with those of a range of other corns.

I built my sand box in order to figure out whether the adaptations of these crop ecotypes were really that unique. In a series of replicated experiments, I compared the Hopi blue flour corn with two closely related blue flours, those from the floodplain soils of Taos and Isleta pueblos in the Rio Grande watershed. I also threw in Golden Cross Bantam sweet corn as a "modern standard." I wanted to test the claim of corn geneticists that modern high-yielders can outperform "unimproved" land races under both optimal and marginal conditions.

How would a Midwestern-bred sweet corn compare with native Southwestern corns when planted in a very marginal "soil": a loamy sand from dunes around Dinnebito Wash below Third Mesa? Highly alkaline, this "soil" was 84 percent pure sand with 7 percent silt and 9 percent clay, certainly not what the Bantam sweets nor the Rio Grande blues were used to experiencing. Finally, I planted all the seeds eight inches deep, watered the sand, and let them go six days.

The differences between the Hopi blue and the Bantam were obvious even before I began crunching numbers on their shoot and root growths. While nearly half of the Hopi kernels emerged in just six days, none of the Bantam seedlings surfaced above the sand. The first leaves of the Bantam seedlings bent and curled beneath the weight of the sand, while the Hopi shoots plowed right up through it. The Rio Grande corns fared a little better than the Bantam, but less than a quarter of them emerged as rapidly as the Hopi.

When my measurements for all four varieties were compared, I discovered that the Hopi blue corn from Third Mesa had both root and shoot elongation rates faster than any of the other corns. The root growth of the Rio Grande blues approached that of the Third Mesa maize, but the latter had much faster shoot expansion—a critical factor if the seed is to emerge from deep plantings. Moreover, the mean shoot growth of the Hopi blue was nearly twice that of the Bantam—9.8 inches versus five inches. Bantam, a widely-*adaptable* sweet corn, was simply not as finely adapted to sand dune conditions as the heirloom Hopi corn. Although their kernels look identical to most observers, the differences in responses between the Rio Grande land races and the Hopi confirmed that "all blues are not created equal." Long ago, the three maizes may have spun off from the same parental stock, but today each is adapted to fit different sets of soil and weather conditions within the heterogeneous Southwestern United States.

IV.

Once natural selection sets a course in native fields, cultural selection and saving of such adapted seeds must be reinforcing. To gain any advantage from them in the long run, farmers must consistently save locally produced seeds rather than always recruiting new stock from another area.

For centuries, local seed-saving was the norm. Ethnobotanist Janis Alcorn has described how traditional farmers follow unwritten "scripts," learned by hand and mouth from their elders, that keep agricultural practices relatively consistent from generation to generation. Most land-based cultures have such scripts that guide plant selection and seed-saving. Each individual farmer might edit this script to fit his or her peculiar farming conditions, but the general scheme is passed on to the farmer's descendants. Thus, the crop traits emerging through natural selection in a given locality are retained or elaborated by recurrent cultural selection.

I first realized the importance of this kind of cultural reinforcement on another sojourn among the Hopi. The Bean Dance again cropped up. I was searching for Hopi lima bean varieties, but had no idea if they were still grown much.

"Oh, yes, I always have some of them stored away," a Hopi farmer replied, "and I grow them every year as well. They are needed in the kiva prior to the Powamu, the Bean Dance in late winter."

"So the *masi hatiqo*, the gray lima, is one of the kinds sprouted for that?" I asked timidly, fearing another indiscretion. I had been to the Bean Dance the year before, and had seen the 12–20-inch-long sprouts delivered by katsina dancers at dawn prior to the dance. The trimmed sprouts and whole corn are cooked in a broth and served that day as a special meal. Bean sprouts are also served to those men in the kiva who have been observing a fast, and refraining from meat and salt as part of the spiritual observances.

The Hopi man laughed merrily. "Yes, that's the one—those *pahaana* [white man] beans won't even work. We had a new boy initiated into the kiva last year who didn't grow out his clan's beans, and thought he could get away with buying some white lima beans at the store to use in the kiva. They were planted in sand just like our Hopi beans, but when our sprouts were tall, his hadn't even come up. Did he ever *hear it* from the other men . . ."

The newly initiated boy was chastised by the other men in the kiva because this late winter underground planting of beans is said to forecast the productivity of the coming crop season. If the beans do well in the sand basins within the kiva, the crops in the summer fields will be plentiful.

Had the boy endangered the coming crop by planting a bean that didn't emerge? Perhaps the spirit of the ceremony was threatened by not paying attention to particulars. Storebought baby limas don't necessarily fit the bill.

This is not to say that the Hopi are purists who don't allow any bean that is not "their type." They often grow exotic vegetables in their faucet-watered dooryard gardens, but those are not the ones that will survive in their most exposed fields or that are suitable for their ceremonies. And even among batches of the true Hopi lima beans there is considerable variation. Part of this heterogeneity may be due to the wide-ranging environmental conditions experienced by their plantings, from dry-farmed dune fields to spring-fed terrace gardens. Levels of nitrogen and soluble salts can vary threefold between different bean plots.

Still, nutritional analyses of Hopi limas indicate a variability that is not only environmental. Genetic heterogeneity in environmental responses must also be taken into account. An eightfold difference in sodium content has been found among Hopi limas. Chemists have also documented a thirty percent difference in crude protein; a threefold difference in iron; and a fourfold difference in crude fat between lima bean types around the mesas.

V.

Hopi farmers are careful to maintain a certain heterogeneity within each cultivated species or crop variety. Or perhaps, I should say, they are careful not to be too careful, or to act as though they could entirely control a crop's destiny.

On one occasion, I asked a Hopi woman at Moenkopi about seed selection for "trueness to type." I had heard that other people discard any unrepresentative seeds in order to maintain a semblance of purity within each seedstock. I wondered if she regularly selected only the biggest kernels, or ones from one end of the cob, or those consistently of the same hue. The elderly woman listened to my loaded questions, then snapped back at me, "It is not a good habit to be too picky. . . . We have been given this corn, small seeds, fat seeds, misshapen seeds, all of them. It would show that we are not thankful for what we have received if we plant certain of our seeds and not others."

Her acceptance of heterogeneity contrasts markedly with the prevailing preoccupations of modern agriculture: uniform seed, for standardized field conditions. Fifteen years ago, genetic resource scientist Jack Harlan warned that negative consequences of this obsession were already being manifested:

"A pure-line mentality, convinced that variation was bad, uniformity was good, and off-types in the field somehow immoral, developed. . . . Thus it was that we laid ourselves open to epiphytotics [plagues] of serious dimensions."

The pest and disease epidemics that sporadically plague modern farmers are most likely to occur where the environment is so uniform that the same seeds can be sown horizon to horizon. The genetic vulnerability of the major crop in a region may be just as much due to this ecological homogeneity as it is to the genetic uniformity of the crop itself. Lacking patches and corridors of other crops as well as natural vegetation to slow down the spread of a pest or disease, these insects or pathogens can rapidly multiply and devastate large areas of one crop genotype.

It may be possible to transform central Illinois into one large cornfield, but fortunately such a monolithic trick is impossible over much of the face of the earth. Environmental inconformities within and between fields are pervasive. It may be too costly and in many cases impossible to smooth off all their remaining lumps and rough edges to suit uniform seedstocks. Because they are resolved to live with multifarious fields, many traditional farmers sow multiline mixtures of the same crop as a kind of insurance against disparate conditions. But it is not only the Hopi, nor Native Americans exclusively, who use the strategy of mixing various crop strains in the same field.

Russian grain farmers, European forage growers, and Latin American manioc planters have also discovered the principle of mixtures.

British scientist B. R. Trenbath has evaluated the productivity of such mixtures, and unhinged the closed door of modern agronomic dogma with his results. Trenbath learned that mixtures of different genetic strains sown in the same field usually yield at least as well as the mean of their components. His work has been confirmed by agronomists and pathologists, who further assert that certain mixtures' yields over several years cumulatively exceed the highest-yielding component strain.

Recently, entomologist Fred Gould has found that mixtures of pest-resistant and non-resistant strains of crops reduce the virulence and evolutionary rates of damaging insect populations. On wheat, it may take Hessian flies less than a decade to overcome a single resistant gene bred into a new variety, particularly if that one variety is sown over large acreages. But if mixtures of half-resistant and half-susceptible wheats were sown together, or if wheat varieties were patchworked in the same agricultural valley, the flies would not do as much damage, nor adapt as quickly to the resistant genes.

Mixing strains of the same crop is one strategy continued by many native farmers, but planting a diversity of species is even more common as a carry-over from prehistoric times. Multiple cropping still accounts for 20 to 40 percent of world food production, and cannot be swept under the rug as an outmoded manner of making a living. Rather, the fact that most land grant agricultural colleges continue to ignore multicropping traditions in no way means that these legacies are less successful than numerous flash-in-the-pan farming schemes that they base on one miracle crop.

VI.

Since 1974, ethnobiologist James Nations has been making yearly visits to Mensabak, Chiapas, Mexico, to document the value of interspecific diversity in Mesoamerican agriculture. His host, known as José Camino Viejo, is a Lacandon Mayan farmer whose plantings lie amidst secondary forest growth within Mexico's humid tropical highlands. There, as Nations writes: "As if in defiance of the politician's declarations that traditional agriculture is obsolete and wasteful, José has farmed the same three-hectare plot (about seven acres) for the past twelve years. A quick survey in 1976 revealed seventy-nine different species of food and fiber crops growing in a single hectare. Light years beyond the Mayanist's holy trinity of corn, beans, and squash, he harvests rice, pineapples, sugarcane, bananas, taro, manioc, yams, limes, spices, oranges, cotton, avocados, rubber, cacao, and pounds and pounds of tobacco, the Lacandones' cash crop."

Between a sizeable tobacco harvest and a maize yield of two tons of shelled seed per acre, José demonstrates a farming strategy that can hold its own in a cash economy. (It has also held its own for more than a decade on one plot, in the midst of a region where quick turnover slash-and-burn is now the norm.) Yet many of his products are for local consumption, not the cash market. They probably go toward meeting the dietary requirements of his family and friends. The considerable variety of vitamins, minerals, essential oils, and amino acids supplied by José's cornucopia of fruits, greens, and grains must go far toward satisfying the nutritional needs of several people.

Victor Toledo, a Mexican ecologist, has noted that biodiversity translates into nutritional variety where cultural knowledge of that diversity is most profound: "If the ecological diversity of the country encloses a rich deposit of nutrients, then it is in the diversity of cultures where one finds the keys to open and use this nutritional treasure."

I sense that James Nations would embrace such a paradigm, for he sees in José Camino Viejo "one of the finest tropical farmers in the Americas and a reminder of the ecological sophistication achieved by native peoples before contact with Western civilization . . ."

Will contact with or conquest by Western civilization bring an end to such sophisticated responses by traditional farmers? The economic and psychological pressures from modern society are now so pervasive—Jolt Cola and mail-order flower catalogs now enter the Mexican barrancas; seed packets of four-year-old (and often inviable) F1 hybrid vegetables are sold in the market at Chichicastenango; sacks of surplus commodity pintos replace the planting of native beans on numerous U.S. Indian reservations; and thinly-clad blondes sell pesticides and hybrids on billboards throughout Latin

America. Is traditional, mixed crop agriculture, as it has functioned for five hundred generations, now doomed?

At one time, after several years of interviewing elderly Indians who were among their cultures' last subsistence farmers, I did assume that traditional agriculture was doomed. I shudder when I remember some of the last River Pima tobacco seeds, buried beside an old farmer before anyone realized that this living gift was the end of the line. After painful deliberation, his family decided to recover the seeds. Although the cache was successfully excavated and germinated, it finally failed to produce the seed crop needed to carry on.

An elderly woman of the *Hia C-ed O'odham* (Sand Papago) culture has used the analogy of a seedstock to describe the fate of her people. She told her relative, Fillman Bell, that:

It was this way, long time ago, when People first realized the world, from that time on it is recognized from their Maker that people who bore children . . . band together. People were like a cultivated field producing after its kind, recognizing its kinship, the seeds remain to continue to produce. Today all the bad times have entered the People, and the People no longer recognize their way of life. The People separated from each other and became few in number. Today all the People (O'odham) are vanishing.

To be sure, there are countless cases of sweeping genetic erosion as acculturation occurs and as agricultural communities are transformed. For years, I worried that the loss of heirloom vegetables, mixed crop farms, and cultural knowledge about them was an inexorable process. I feared that the genetic erosion of crops was globally outstripping rescue efforts, possibly imperiling our species' capacity to feed itself in the future. I also lamented the extinction of two hundred or so Native American languages in Mexico, Canada, and the U.S. since European contact intensified five hundred years ago, for it meant the demise of much culturally-encoded knowledge about these crops and methods of growing them.

Genetic erosion has certainly wasted a considerable wealth that once existed on the American continents, but I no longer believe that it will sweep away those remaining traditional seedstocks that have so far escaped its reach. The economic and psychological forces that aggravate crop genetic erosion and disrupt traditional agriculture simply do not penetrate all places and all cultures equally. After hundreds of years of economic or military domination by colonial powers, there remain agrarian peoples who continue to grow many of the same seedstocks their forebears did centuries ago. There are others who farm in the shadows of multi-million-dollar plant breeding centers, yet choose to plant their family's heirloom seeds, making their own selections and mixes from them, rather than buy hybrid seeds.

This realization was suddenly brought home to me when I visited the International Maize and Wheat Improvement Center (CIMMYT) at El Batán in the state of Mexico. For more than two decades, CIMMYT's breeders have been releasing hybrid corn cultivars on a worldwide basis, dramatically improving the grain yields of many developing countries. And yet, when the center's economists began to assess the impact of this intensive international development effort, they learned that over half of all the corn planted remains open-pollinated seed saved from local harvests. Despite half a century of pervasive advertising and public education encouraging corn growers to obtain improved, hybrid seed from off-farm sources, the majority of them either grow family heirlooms or obtain some of the seed of an introduced variety from a neighbor who has demonstrated its utility under local conditions.

Driving out of CIMMYT headquarters toward the nearby college town of Texcoco, we saw numerous fields where maize and squashes were interplanted, with runner bean patches on the margins. A corn specialist with us commented offhandedly, "That's mostly Chalqueño, a native variety that's been in this area for a few hundred years. . ."

"*What?*" one of the other passengers blurted out, flabbergasted that it wasn't a CIMMYT variety.

"Well, with Chalqueño, those farmers are achieving yields as high as ten to eleven tons to the acre. . . . Yields that high are respectable even in the U.S. Corn Belt. . . . You just can't do much better than that here, given the level of inputs invested in these crops. There's hardly an incentive to obtain improved seed under these conditions . . ."

CIMMYT corns and wheats have found a home in countless other places, so that "hometown acceptance" should not be the only criterion used to judge its success. Yet as early as the forties, applied anthropologists began to indicate to breeders that their hybrids would not be adopted everywhere, even though they may indeed be higher-yielding under most conditions.

In New Mexico, Hispanic farmers were initially interested in the higher-yielding corns provided to them by extension agents. Within two years, however, many of them returned to growing their native corn, as if the introduction of the hybrids had never happened. When extension agents inquired about this reversal, the farmers explained that their wives liked neither the taste nor the texture of the introduced corn, and refused to use it for making tortillas. One by one, the farmers returned to growing the corns favored by their wives. To this day over much of New Mexico, blue Pueblo flours and flints are grown for tortillas and cornmeal, and Mexican Junes for white hominy, while hybrids are preferred only where corn is grown for animal feed.

The more marginal the land is in arability, the greater the probability that native crops will retain their advantages over introductions. Moving from the fertile valley floor to the upland perimeters of a watershed in the Andes, anthropologist Stephen Brush has observed marked differences in the kinds of crops that farmers grow. In central Peru, improved potato varieties have been introduced into the highlands since the early 1950s. In the lower reaches of the Mantaro Valley, near major urban and market centers, researchers found that 87 percent of all fields were planted to improved varieties. But above ten thousand feet on the valley slopes, roughly half of the fields remained in native potatoes. Beyond twelve thousand feet, in a zone that no North Americans would dare to cultivate, Andean farmers successfully plant several kinds of tubers, potatoes among them, with 88 percent of the fields seeded to native varieties.

Faced with formidable climatic constraints, the high-elevation farmers switch to frost-resistant bitter potatoes. Called *shiri-akshu* in the Quechuan tongue, these spuds are freeze-dried into a palatable foodstuff, *chuño*, and the bitter glycoalkaloids are removed in the process. The *shiri-akshu* varieties divide into two species that are rarely dealt with in modern breeding programs. Brush has noticed that even where the highland farmers provisionally try improved varieties, they still keep one or more of their field patches reserved for mixed native varieties alone.

Ultimately, Brush concludes, Andean farmers have different reasons for growing native and improved varieties; both may have a place. Newer varieties are considered watery and poor tasting in traditional dishes, and are liked only by "undiscriminating urban consumers." They also degenerate under storage and cannot be saved long as seed potatoes. However, the improved cultivars of the common potato have been selected for frost and blight resistance and responsiveness on valley soils. They therefore yield well as a cash crop to market in the cities where those undiscriminating consumers live.

On the other hand, the "floury" quality of native varieties, termed *machka* in Quechua, is of prime importance to native farmers who typically eat potatoes as their primary food two or three times daily. The easily-stored native potato selections, each unique in taste, texture, and color, add variety to the Quechuan diet in a way that a Russet cannot. There is little cause for worry over the wholesale replacement of forty to fifty native varieties by a single high-yielder, at least in potato patches above 10,000 feet. In the central highlands of Peru, Andean weather and indigenous cultures have been partners in selecting potatoes for 5800 years.

Ironically, it is not the International Potato Center in Lima, but a monolingual Quechuan speaker who has been most successful in selecting new potato varieties for high-altitude environments. Over the last few decades, Eugenio Aucapuma has become a major source of good-tasting, climatically adapted dark potatoes, developed through backyard selection. Christine Franquemont has documented how one of Aucapuma's folk varieties, known as "papa olones," has spread from Mr. Aucapuma's patch of mixed land races, wild species, and family selections. Because Aucapuma was selecting for taste, color, and hardiness—all qualities that most Quechuan farmers value in potatoes—"papa olones" rapidly passed from his patch to his neighbors', and then from the Olones valley to similar sites throughout the mountains of southwestern Peru.

It is clear that such "traditional farmers" are not simply resistant to change or recalcitrant when it comes to accepting any fitting seedstock or technology from outside their own culture. Instead, they are looking for qualities that are not usually found in the foods and seeds and breeds emanating from modern breeding programs. They are selective, accepting only those items from other cultures that may have lasting value in their own.

And here, at last, is the common thread that weaves together these stories from the Hopi, the highland Maya, and the Quechua: their native agricultures continue today because they are persistent cultures, retaining sets of values not found in the modern marketplace. The late anthropologist Edward Spicer has argued that these enduring cultures, however marginal they may look from New York or Rome, "help to determine the character of civilizations by asserting a quality of the human spirit not typical of dominant peoples." Spicer was among the first applied anthropologists to articulate what qualities these persistent cultures share. His ideas also provide much food for thought regarding enduring agricultural communities.

It was abundantly clear to him that hundreds of Native American cultures have fallen by the wayside, ravaged by battle, disease, or colonial dominance; and that whatever agricultural wisdom they held has been muted. Still, a relative few have had the tenacity to keep their native crop gene pools, agricultural rituals, and land-based symbols intact. Regardless of their intensity of contact with outsiders, invaders, or foreign powers, their ethnicity or *other-ness* has been retained. For farming peoples, this ethnicity can be read in the colors of their seeds, and in the patchwork designs of their fields.

But why, Spicer asked, have some peoples succumbed to acculturation and assimilation, while others have endured culturally? When he began to compare the cultural traits of nine enduring ethnic groups, Spicer quickly found that easy assumptions—that the "keepers" forbid intermarriage with other cultures, or defended their homeland against all odds, or maintained a language to encode secrets indecipherable to outsiders—all fell by the wayside. Most of the enduring peoples had lost some of their genetic identity, some of their original territory, and, often, much of their native language.

Despite these losses, something basic has prevailed. Certain tribes such as the Cherokee have become racially intermixed, but they themselves can still tell who upholds traditional values. And at least in North Carolina, these traditional Cherokee maintain corn varieties quite unlike those found anywhere else in the Eastern U.S. today. Spicer also learned from his friendships with the Arizona Yaqui that a century after fleeing from their original Mexican homeland, they remained in contact with relatives who still lived in the Sonoran pueblos. Seeds such as tepary beans, chiles, and bottlegourds are still traded up from their ancient territory. More importantly, they maintained a symbolic relationship with their original villages, and with the legendary Talking Tree of their mother country. According to Spicer, fewer than a third of the Seneca still speak their native language, and yet most of them still counterpoise their own identity against that of Anglo-Americans by using indigenous terms: *okwe'ooweh* versus *hanyo'o* 'whites'. They continue to refer to the Code of Handsome Lake, and have been among the last of the northeastern tribes to cultivate native crops such as sunflowers and colored corn. Feasts of corn and wild strawberries remain associated with recitations of the *Gai'wiio*, the prophesy of Handsome Lake, which has been passed down by word of mouth through various authorized "holders" of the prophesy among the Seneca and the other Iroquois tribes of the Six Nations.

What counted, Spicer found, was that if enough members of a community express an affinity to shared symbols and values through time, across the generations, their culture will remain viable. This common covenant must be constantly transmitted and reinforced by the legends, codes of behavior, ceremonial songs and dances, rituals, vigils, and homages to places sacred to that particular culture. Once inoculated with the values held within their community, children acquired a certain resistance to the trappings of the dominant culture. Perhaps these values act as "cultural antibodies," which react to outside influences in such a way as to overcome their negative effects.

And yet, another metaphor has recently emerged that more deeply describes the manner in which persistent peoples live within a place. They manage local resources in a way that is informed by what ethnographer Eugene Anderson calls "an ecology of the heart." Anderson refers to the emotions and meanings that guide traditional cultures in their relations with the places, natural processes, and organisms around them. As Apache Nick Thompson told Keith Basso, "If you live wrong, you will hear the names and see the places in your mind. They keep on stalking you, even if you go across oceans. . . . They make you remember how to live right, so you want to replace yourself again."

Ultimately, it is this ecology of the heart that keeps a Hopi lima bean planted in sand dunes, allows a Lacandon garden to be rich with dozens of tropical food plants, or lets an unknown blue tuber emerge from a Quechuan potato patch. "Without an intense, warm, caring, emotional regard for the natural world," Anderson reminds us, "we will be literally incapable of preserving it."

And that is the crux. If we are to have any quality in our home environment, in our food, and in our farming, we must seek to embody enduring values as well. Anderson reminds us:

"We are paying a dreadful price in species extinctions, deforestation, soil erosion, agricultural degradation, and the rest. *Human beings make sacrifices for what they love*, not for what they regard as merely a rational means to a material end."

A cultural community that persists in its farming tradition does not simply conserve indigenous seedstocks because of economic justifications; the seeds themselves become symbols, reflections of the people's own spiritual and aesthetic identity, and of the land that has shaped them. Anderson believes that these traditional communities have ". . . found ways to involve not only love but the whole panoply of human emotions in the conservation effort. We must learn from them or we will not survive."

How we care for the lima bean sprout—or the misshapen maize kernel and the bitter potato—may indeed forecast our own survival.

CHAPTER SIX

New and Old
Ways of Saving:

Botanical Gardens, Seed Banks,
Heritage Farms, and Biosphere
Reserves

I.

Not long ago in Central Mexico, I unknowingly walked through the site of a pre-Columbian botanical garden, casually observing flowers and tallying the plant species. The vegetation on the garden grounds hardly offered an indication that a distinctive flora had once been intentionally planted there. When I later learned of certain Aztec codices, and read accounts by early Spanish naturalists in the New World, an altogether different picture emerged.

Around 1450 A.D., the Valley of Mexico was known for its gardens of earthly delights. These were not the orchard gardens of mixed fruits and vegetables, nor the commonplace floral beds that the Aztecs called *Xochi-chinancalli*, 'flower place enclosed by reeds.' Instead, I refer to enormous pleasure gardens, some of them two leagues in circumference. There, the Aztec elite would come to bathe in hand-sculpted pools, to listen to the song-birds, and to watch water cascading down the fountains. In an atmosphere of mist and color, they would become intoxicated by the perfume of exotic flowers, and, perhaps, by the juices of psychotropic plants.

Imagine these gardens as artificial tropical rainforests, or as massive collections of medicinal, hallucinatory, and sweet-smelling plants. Many of these plants were introduced to the Valley of Mexico from the tropical coasts hundreds of miles away, delivered to the gardens in huge clay ollas by crews of slaves. The arboreta of the Aztec emperor, Montezuma, were described by historian Cervantes de Salazar as

> . . . spacious gardens with paths and channels for irrigation. These gardens contain only medicinal and aromatic herbs, flowers, native roses, and trees with fragrant blossoms, of which there are many kinds. He ordered his physicians to make experiments with the medicinal herbs and to employ those best known and tried as remedies in healing the ills of the lords of his court. The gardens gave pleasure to all who visited them on account of the flowers and roses they contained, and of the fragrance they gave forth, especially in the mornings and evenings. . . . In these flower gardens, Montezuma did not allow any vegetables to be grown, saying that it was not kingly to cultivate plants for utility or profit in his pleasance. He said that vegetable gardens and orchards were for slaves or merchants.

The goal was to weed out utilitarian concerns and launch the leisure class into a sublime experience by inundating them with sensory stimulants. Particularly impressive was the Tetzcotzinco garden of Netzahualcoyotl, poet-king of Texcoco. This garden was first described in 1530 by the king's descendant, Ixtilxochitl: "Of the arboreta, the one most pleasant yet curious was the forest of Tetzcotzinco, because . . . it was like falling into a garden raining with aromatic tropical flowers . . ."

Like the other gardens that Netzahualcoyotl designed between 1426 and 1474, this one contained "sumptuous palaces beside fountains, canals, drains, tanks, baths, and other intricate waterworks . . . planted with many strange and wonderful varieties of flowers and trees, *brought thither from remote places.*"

When I first visited Tetzcotzinco five years ago, I could trace the outlines of the former fountains, walk the walls of terraces, and ponder the foundations of temples, but the exotic horticultural heritage had vanished. In fact, I was introduced to Tetzcotzinco while collecting a weedy bean, *Phaseolus anisotrichos*, which is rather common to secondary forest vegetation and roadsides throughout Mexico. It was not that Tetzcotzinco was the only place where these beans could be found. The mountain was simply within easy access of a bean breeder's office in Texcoco, and the ancient stairways enabled a quick ascent of the slopes. The bean breeder briefly explained to me the archaeological import of the mountain's shrines, but failed to elaborate on its pivotal position in the history of New World gardens. I took it for an overgrown architectural monument, but had no immediate sense of its horticultural legacy.

As I later learned, botanists have found on its grounds only one plant reminiscent of earlier introductions from five centuries before: *zapote blanco*, a fragrant fruit tree. As a long-term reservoir of unusual plants, Tetzcotzinco has failed. Of course, that was not its explicit purpose. It was designed to indulge the aristocrats of its era with an air of the supernatural. But both the Aztec empire and its exotic plants met with mortality far sooner than Netzahualcoyotl could have ever imagined.

II.

Recently, I visited Tetzcotzinco again, immediately following meetings hosted by CIMMYT, the maize and wheat breeding center popularly known for fomenting the "Green Revolution." Our discussions had centered on the problem of genetic vulnerability in crops, but had touched on the roles that seed banks, botanical gardens, and crop introduction stations could play in alleviating genetic erosion. During the day, our group had visited CIMMYT's vault of cold-stored grains, and then, in the evening, we attended a reception on the flanks of Tetzcotzinco, with geneticists, botanists, and breeders from a dozen nations. As our hosts unfolded the history of the gardens to their foreign visitors, we looked out over the ancient terraces at dusk. As the darkness gathered, one question kept turning over and over again in my mind: Would the seedstocks down in the gene bank at CIM-MYT last any longer than the species that the Aztecs planted on the slopes above us?

The issue raised by this question has become the focus of a bitter debate among plant conservationists. It pits the *in situ* conservation of plants in their natural habitats and cultural contexts against *ex situ* conservation of seeds frozen in liquid nitrogen or isolated in garden plots and petri dishes far from their origin.

Each of these two approaches to plant conservation has dramatically different advantages and disadvantages associated with it. When I visited him in Washington, D.C., several years ago, science policy analyst Bob Grossman had begun to articulate some of the key trade-offs that result from choosing one or the other of these strategies. Collections in botanical gardens and seed banks are favored by most geneticists, because these resources are usually so accessible for their breeding programs. A breeder can phone up a computerized data base, scan it for the plant materials he wants, order and receive them within a matter of days, unless political embargoes or phytosanitary quarantines interfere. Generally, the overall number of species that can be saved in these *ex situ* gene banks is low, but the number of varieties of any particular economic species is high. Grossman suggests that the storage and growout costs for each species is high, so that only the most useful food, forage, and forestry crops are usually maintained.

On the other hand, *in situ* conservation of wild plant genetic resources today is accomplished largely as a side benefit of protecting scenic wildlands. Within their natural habitats, a large number of species, both useful and apparently useless, can be conserved and continue to evolve. They will continue to function in ecosystems where, inevitably, a few species will be lost owing to "background extinctions." In these ecosystems, however, the overall cost of conserving many species for more than two hundred years' time will be low. Nonetheless, as Grossman told me, crop breeders do not consider plants kept in natural areas to be readily available to them. In fact, the very presence of potentially useful species may go undocumented in a natural area. Two Canadian conservationists, Robert and Christine Prescott-Allen, are critical of Third World parks and reserves where they feel that crop relatives are "highly vulnerable to the pressures of population growth, poverty, and development," even though few parks anywhere have been explicitly managed for their genetic resources for any length of time.

Grossman and the Prescott-Allens contend that if we are to better safeguard plant resources for future generations, *in situ* and *ex situ* conservation strategies must be used in a complementary manner. If we are to pose as the stewards for the remaining richness of the living world, we must draw upon both approaches, so that the weaknesses of the one can be checked by the strengths of the other.

What worries me about current efforts is that few modern conservationists are aware of the precedents, the past successes and failures, of the particular strategies now being promoted as panaceas. Seed saving in America did not start with the building of the National Seed Storage Laboratory in Fort Collins in 1958, nor were rare plant sanctuaries unknown before The Nature Conservancy was founded in 1951. Certain conservation practices predate the written record of European arrival on this continent. If we wish to conserve useful plant species for posterity, we will do well to learn the risks that are involved with every available strategy, none of which are foolproof or flawless. We can only hope that our plant conservation efforts remain effective as long as the *zapote blanco* has survived at Teztcotzinco, but we must not forget the hundreds of other species that were extirpated there during the last five hundred years.

III.

The long-term persistence of plants in botanical gardens does, in fact, have some precedents. The propagation of *Franklinia altamaha* at Bartram's Garden is one well-known success story. Native to the southeastern U.S., *Franklinia* was originally collected by John Bartram on his first botanical exploration as a "Royal Botanist" for George III in the late 1760s. John Bartram and his son William later realized that this tree with pure white blossoms had been seen in only one location, where it covered less than three acres along the Altamaha River in Georgia. Around 1776, William Bartram brought propagation material of the plant back to their hundred and twelve acre farm on the Schuylkill River, where it has remained in cultivation continuously to this day.

The Bartram farm's rural setting has disappeared over the centuries; the southwestern section of Philadelphia has grown up around its remaining forty-four acres. Both John and William managed a portion of it as a nursery, as did their descendants until 1850. While it has recently required restoration to better reflect an eighteenth-century landscape, Bartram's Garden is arguably the oldest botanical garden and living historic farm in the United States. From its grounds, John and William sent seed and nursery stock to other botanical institutions around the world, many of which have continued to grow plants from the Bartram legacy.

Franklinia altamaha is among these horticultural survivors, and it is now featured at several botanical gardens and arboreta. During a visit to the Arnold Arboretum in Massachusetts, I asked the Center for Plant Conservation staff to take me to see this tree. We made our way up a knoll, and stood silent before it, as if asking for a blessing. I might as well have been prone, on a prayer rug, facing Mecca.

However responsive *Franklinia* has been in cultivation, it has not fared as well in its natural habitat, where it was last seen in 1803. Shortly thereafter, the only known example left in the wild was felled when the land was cleared for agriculture. If it had not been for the Bartrams' earlier collections, the species would now be extinct. Instead, should a site in the Altamaha watershed ever be secured, the remnants of this species could be reintroduced to the wild. The Center for Plant Conservation is now involved in several other reintroductions, for plants that were incidentally collected years ago, then lost in the wild, and which are now being transplanted into their former habitats.

Unfortunately, *ex situ* sanctuaries for rare plants don't necessarily receive the support they need, even when they alone are keeping species from extinction. This became painfully evident to me when I learned the brief history of the Rare Plant Study Center. This garden and greenhouse center for endangered plants was established at the University of Texas in 1971 by Dr. Marshall Johnston and other farsighted botanists.

One of the Center's first efforts was the rescue of *Manihot walkerae*, a relative of cassava already extirpated in Texas and close to extinction in the neighboring Mexican state of Tamaulipas. It also salvaged the wild Texas pistachio before the Amistad Reservoir on the Rio Grande inundated the last population of that species.

"The pistachio was already listed as extinct in North America," as Marshall Johnston remembers. "When we heard there was still this one site, and that it was about to be destroyed, we hurried down. We had to go after it in a boat."

Fifteen years later, Dr. Johnston led me to a little patch of plants, including the wild cassava and pistachio, in front of the botany building on the Austin campus. The Center's modest financial support had long since diminished, as scientific funding fads skipped from ecosystem modeling, through chemotaxonomy and coevolution, to molecular genetics and biotechnology. No longer receiving adequate support for his rare plant studies, Johnston simply keeps an eye on these plants, so that the grounds maintenance crews on campus do not destroy them.

A few more fortunate botanical gardens have secured stable financial backing for their conservation efforts. Nonetheless, certain kinds of endangered plants are prone to vanish from their living collections. Despite their many skilled horticulturists, botanical gardens frequently undergo changes in direction and priorities, leaving some kinds of plants to be neglected. Because they often cater to social fads in promoting exotic ornamentals, botanical gardens have chronically ignored obscure native species that aren't particularly showy. The accidental introduction of a plant disease, a blunder by a groundskeeper or visitor, or a catastrophic freeze can swiftly eliminate a propagated plant that is scarce in numbers to begin with.

Dr. Thomas Elias of Rancho Santa Ana Botanic Garden is quick to note how many things can go wrong when a rare plant population is kept in propagation for many decades. As one of the editors of the widely acclaimed anthology, *Extinction Is Forever*, Elias has seen enough plant conservation projects to become acutely aware of their typical limitations, as well as of their strengths. When I invited him to speak to propagators of endangered desert plants, I knew he believed that gardeners can play an effective role in conserving endangered plants if they are diligent in maintaining the population size of each distinct field collection.

Dr. Elias has gathered data on eighty-five rare species that have successfully been grown in botanical gardens, but his most sobering case study is that of the rare Nevada variety of a species known as the *Agave utahensis*. This small century plant is akin to crops such as tequila, sisal, and henequen. Rancho Santa Ana first received 204 of these plants in 1936, a sample theoretically large enough to maintain genetic heterogeneity for decades, if not centuries. And yet, Elias discovered that the number of plants in the sample had dramatically declined by 1951, when only "fifty of the original plants were still alive. During that year, Rancho Santa Ana Botanic Garden moved . . . to new facilities in Claremont, California. An undisclosed number of this introduction was moved to Claremont and by the 24th of May, 1981, only sixteen plants had survived." From the two hundred-plus originals, the vegetative offspring of no more than sixteen plants persist today. These progeny probably represent only a meager proportion of the genetic variation that was in the original field population.

Another California garden director, Dr. Harold Koopowitz, points out that inbreeding depression may take its toll when population samples dwindle to this size. After generations in cultivation with no gene renewal from wild plants, some ornamental flowers that are normally self-pollinated have lost their ability to produce stamens and pollen. Others, Koopowitz observed, have changed their flower colors or shape to the extent that they no longer attract the pollinators that served them in the wild. Without vegetative propagation or artificial pollination, these horticultural plants can easily become evolutionary dead-ends, unable to survive if returned to the wild. Horticulturists must consciously work to avoid such problems, or their conservation efforts will not bear fruit for future generations.

IV.

Frozen seed banks were intended to head such problems off at the pass. Not long ago, the United States Department of Agriculture realized that of the 160,000 crop samples or "accessions" that had been delivered to its plant introduction stations since 1898, only five to ten percent were still alive and accessible. Some of the seed collections could not be grown out immediately, and soon lost their viability. Others, planted every few years to keep them viable, had been lost altogether due to crop failure, contaminated through un-

controlled cross-pollination, or suffered the detrimental genetic changes
that Koopowitz has described. The obvious alternative was to keep their
seeds viable for longer periods of time, thereby reducing the frequency of
growout, and the chances of crop failure, contamination, or genetic drift.

During the twenties, Freon-based refrigerators became available to plant
breeders who, up until that time, had only been storing their seeds at room
temperature. Refrigeration can slow down the deterioration of seeds for
many crops and may also reduce their exposure to insect infestation. The
USDA began switching over to cold storage of seeds in the forties and fifties,
keeping their supplies in metal cylinders filed within cold vaults not unlike
meat lockers. Today, these same seed banks are transferring their holdings to
humidity-free sealed envelopes, stored at minus ten degrees centigrade, or
lower. Some scientists have even switched to immersing freeze-resistant
plastic tubes of seeds in vats of liquid nitrogen.

At the National Seed Storage Laboratory in Fort Collins, Colorado, Dr.
Eric Roos led me past two enormous vaults packed full of stainless steel cylinders of seeds. He opened the door of a small room that had a waist-high
canister on the floor. Putting on a white lab coat, and a pair of "cryo-gloves,"
Eric opened the port of the liquid nitrogen storage reservoir. I thought he
would expose the ton of seeds and chemicals inside, sitting in the crispness
of minus 196 degrees centigrade. Instead, a cold mist spilled out of the tank,
because most of the liquid nitrogen was in vapor phase, at a "mere" minus
150 degrees. Eric hooked a tube hanging on the lazy-Susan rack inside, and
pulled it from the mist. As Eric explained the advantages of this kind of cry-
ogenic storage, I stared at the thousands of frozen seeds packed into the tube
he held in his gloved hand.

Such nitrogen reservoirs can store seed accessions in a fraction of the
space they would require in the vaults down the hall from us. They are not
so vulnerable to power failures as are the electrically-run Freon refrigerators
that cool the vaults. Each foot-long tube can contain as many as fifteen thousand potato seeds, while tens of thousands of tubes can be packed into a single reservoir. Already, seeds of more than a hundred and fifty species, from
Native American sunflowers to saguaro cactus, have weathered such deep-
freezing, still able to germinate.

Seeds stored this way could hypothetically last longer than we have so far lasted, *as a species*. If Dr. Eric Roos and his colleague Philip Stanwood are correct, seeds in a deep-freeze should suffer no more than a one percent loss in germination over a hundred billion years.

And yet, the strategy of putting seeds into cold storage as a back-up conservation measure is actually centuries old in the Americas. I learned this in a village on the U.S.–Mexico border, talking with a Tohono O'odham couple. Their ancestors had a long tradition of sealing seeds within pots to be stored in caves near their fields. Such a practice appears mundane at first glance, but actually profound in its understanding of seed longevity and the unpredictability of the desert.

In years when better-than-average rains offered O'odham farmers a surplus at harvest time, much of it would be shared with surrounding communities that may not have been so fortunate that season. As a back-up, some of the surplus seedstock would be placed in a particular kind of clay vessel called a "seed pot." A seed pot would be shaped with a narrowed neck so that a lid could be wedged into its mouth and glued shut with a lac gathered from creosotebush stems. These pots were sequestered in caves and rock crevices.

The O'odham woman used the term *kaicka* to refer to seeds specially reserved for future plantings, as opposed to *kaij*, which may refer to any seeds, even those that may be eaten. Her husband told me how his forebears had put some of those pottery jars away years ago, hoping to come back to them when needed. But he can't leave the pots around anymore.

"One of our pottery jars was taken away not long ago. Then a few years later, we were watching TV one Sunday morning, and they showed some Indian pottery on there. Sure enough, it was our family's clay *olla* full of seeds. We called the TV station, but the one who talked to us said he didn't know where that picture was taken. We never did see the jars or the seeds ever again."

Placed in the relative coolness and constant dryness of a desert cave, most kinds of crop seeds grown by the O'odham can remain viable for a decade or more. The desert crop seeds, because they were hermetically sealed into fired pottery jars with the lac-covered lids, were also protected from insects and rodents. This offered farmers a reserve of seeds in case several years of droughts, floods, or other stresses left their planting stock for the coming year in short supply.

Some native farmers don't necessarily plant the same kind of corn every season, but vary their selection depending upon the weather. By having caches of other seedstocks for particular weather conditions, farmers can hedge their bets. Anthropologist Tim Dunnigan has told me of this kind of crop switching among Mexico's Mountain Pima, relatives of the Pima and Tohono O'odham of Arizona. They keep several varieties of maize on hand, including a quick-maturing corn in case spring drought should delay planting, thus shortening the growing season. Other Mountain Pima maize varieties can be planted earlier in cool, wet springs, but need five months to mature. Higher in the Sierra Madre Occidental, the Mountain Pima and their neighbors sometimes use caves as granaries, keeping their seed harvest in cold storage for much of the year.

In the early eighties, genetic conservation specialists returned to the idea of storing seeds in caves. José Esquinas-Alcazar of the U.N. Food and Agriculture Organization considers a proposal for storing seeds in large caves high in the Andes more cost-effective than the use of conventional seed banks. If located away from fault zones in an alpine area sunny enough to allow the use of photovoltaic cells for lighting and instruments in staff work areas, such a montane deep freeze might also be safer from catastrophes, blackouts, and brief political or economic crises. Andean cave storage, though surely effective, simply has not captured the imaginations of the high-tech engineers who usually design gene banks. To date, this alternative has also not captured the financial backing of the foundations and agencies which support most genetic conservation work.

Nonetheless, the notion of putting a freeze on genetic erosion—whether it be with liquid nitrogen, Freon, or alpine caves—has its own inherent limitations. By removing your plants from frequent contact with field conditions, and "locking them up" for decades at a time, you could put their evolution in cold storage, on hold. As agricultural environments change outside the liquid nitrogen vat—as pests and diseases rapidly evolve new virulence—the stored crops are not coevolving with them. As a result, they have no chance to develop the needed resistance.

I can imagine a scenario in which the rare, valuable plants now sheltered in seed banks, once released, succumb to newly-arisen problems that did not exist when they were put into storage. Consider the demand for salt-tolerant plants that will exist by the year 2089, when conventional crops will fail on much of the world's desert farmlands that have been salinized by poor quality groundwater. Let us imagine that at that time, a crop breeder searching for salt-tolerant plants discovers that a twentieth-century scientist once banked seeds of a wild bean relative that grew on saline playas of the Mexican coast. That wild bean population is now extinct in the wild, and no other sources of salinity tolerance are known for cultivated beans. He summons this sole accession up from the National Liquid Nitrogen Pool, where it has been stored for one hundred years. It arrives in the mail the next day.

The scientist sprouts it in a peat pot in the greenhouse where he normally hybridizes his beans. Instantly, most of the plants become infected with a whitefly-transmitted virus strain that didn't even exist when the seeds were first stuffed into a tube and dunked in liquid nitrogen. The whitefly has evolved to resist the most frequently used greenhouse insecticides, and the virus itself has become more virulent as the greenhouse industry has expanded. Worried, the plant breeder removes the surviving plants from the greenhouse, and attempts to grow them outside in a small garden at his Eastern university. There, they fail too, overcome by air pollution, or by soil contamination from acid rain. The salt-tolerant genes have been rendered meaningless by the altered conditions they find on the ground a century later.

As a kind of survival insurance, seed banks may be fine, but there will be tremendous losses if we assume that they are all we need in the way of long-term conservation measures. Frozen banks may store germplasm, but they cannot ensure that the value of this germplasm will be recognized and used. For one thing, many of the collections now in gene banks have little more than an accession number and a country of origin attached to them as their passport or "pedigree papers." Looking at this passport data, one can seldom tell whether or not an accession has salt-tolerant genes in it, even though it may have been collected from saline playas; such information is almost always lacking. Hundreds of thousands of such accessions have been inadequately documented when first collected for the USDA, largely because its plant exploration program has been so meagerly supported that most samples were grabbed in a hit-and-run fashion. Particular habitats, uses, ceremonies, or locally-perceived attributes associated with these plants are typically not noted by one-shot collectors.

When such information fails to be recorded, these seeds lose part of their value. Like wheat germ with the bran cracked off, they have lost their taste and texture. Their environment, and the folk knowledge which once surrounded them, has been left behind, in most cases forever.

V.

In the winter of 1975, Kent Whealy became concerned with the high rate of divorce between crop seeds and farming folklore. As he modestly tells it, his involvement with heirloom seeds and folk varieties of vegetables and flowers began "by chance." I hesitate to agree that family ties and folklore work completely by chance; they work deeply within us, and when action suddenly emerges from us, we often mistake it as a "chance event." Even so, Kent's efforts began humbly, as he has explained them to me and to many others:

"In the early seventies, my wife and I were living in the northeast corner of Iowa. We were newly married and planting our first garden together. Diane's grandfather, an old fellow named Baptist Ott, took a liking to us and was teaching us some of his gardening techniques. He was a gruff old fellow and one of the best storytellers I have ever met. We became really close.

"That fall, he gave me the seeds of three garden plants that his family had brought over from Bavaria four generations ago—a large, pink German tomato; a small, delicately beautiful morning glory, purple with a red star on its throat; and a prolific climbing runner bean. The old man didn't make it through the winter, and I realized that if his seeds were to survive, it was up to me.

"At that point, I had no idea how prevalent heirloom varieties were, but I knew from Baptist Ott's garden that the ones I'd been given were excellent. I began to wonder how many other people might be keeping varieties that had been passed on to them by their families. . . . I also wondered how often elderly gardeners passed away leaving the seeds they had been breeding for a lifetime sitting in a can on a shelf somewhere until they died too. I decided to write letters to the gardening and back-to-the-land magazines to see if I could locate other gardeners who were keeping heirloom seeds."

Initially, Kent was rather disappointed at the lack of response to his proposal, the beginnings of the Seed Savers Exchange. That lack of response, and his initial disappointment, have long since vanished. His members have now made an estimated half million plantings, supplying to one another close to five thousand heirloom seedstocks that weren't in any seed catalog at the time and in some cases were on the verge of extinction.

But it was not just the seeds that were close to vanishing; the traditional gardening techniques, the backyard harvesting practices and food preparation methods associated with these vegetables were all on their way out too. We were losing the stories about the origins of these crops, and the origins of their colorful names: Pruden's Purple (Potato-Top) Tomato; Stove Wood Bean; Blue Eagle Corn; the Moon and Stars Watermelon.

Take the case of another watermelon, "the Nancy," recently described in the Harvest Edition of the Seed Savers Exchange. A white-seeded, red-fleshed, pinstriped melon was discovered by one Nancy Tate in the 1880s, while she was out hoeing a Georgia cotton field. She saved the seeds from this sweet, disease-resistant melon and shared them with neighbors in the Rome, Georgia, area. Later, her cousin took some seeds of the Nancy with him to Arkansas, where it helped to initiate commercial melon farming in an area historically recognized as the watermelon capital of the U.S.

As commercial watermelon growers switched to hybrids that had thicker, more shippable rinds than the Nancy, this heirloom variety fell out of commercial use. The progeny of Nancy Tate's cousin, D. J. Wofford, Jr., and Wade Roy Wofford, kept it growing in Arkansas until the fifties, when they too lost it. Then, in 1986, as Wade Roy Wofford recently reported to Kent Whealy, "my father was able to regain the Nancy for us through 78-year-old Avery Bryan, Nancy Tate's sole surviving son."

When I first went to meet Kent, I knew that he was already in hot pursuit of many heirlooms such as the Nancy, and the Moon and Stars melons. I remember sitting on a Greyhound bus on my way to the Whealy homestead outside of Princeton, Missouri, in 1978. Next to me was a Mennonite farmer from Kansas, and we struck up a conversation.

"What do you grow on your farm?" I queried.

"Beans."

"Beans!" I exclaimed. "Oh, I'm going to visit a family outside of Princeton that grows all kinds of beans: Cornfield Snap Beans, Cliff Dweller's Beans, Bird's Egg Pole Beans, Cherokee Shellouts, Lazy Housewife, and Missouri Wonder. Beans like that. What kinds do you grow?"

The Mennonite farmer looked at me for a moment.

"When I said I grew beans," he explained with a tired look on his face, "I meant soybeans . . ."

If that farmer could have seen Kent's collection, perhaps he would have caught the excitement that my blundering acclamation could not convey. Kent and Diane had already recovered through the mail hundreds of varieties of vegetables from all over the country. Many were kinds no longer available commercially, and were unknown to seed banks and agricultural museums.

Even so, this pursuit was simply Kent's avocation at the time; he had to work as a printer to keep his family and this hobby afloat. His homestead was not well suited for growing out the many vegetables he had received, and the mail was overwhelming him in his "spare time."

Today, a decade later, the Whealys are finally situated on fifty-seven acres of rolling land northeast of Decorah, Iowa. Called the Heritage Farm of the Seed Savers Exchange, it is a fitting place for saving seeds that are cultural heirlooms. Not far from the Heritage Farm, Amish, Finnish, Bavarian, Hmong, Norwegian, and German Catholic communities continue with many of their rural traditions. And Kent, along with several grant-supported colleagues, is pursuing the conservation of our country's crop heritage full time.

Is it *in situ* or *ex situ* conservation? It may be considered a little of both. Kent and Diane still grow the pink German tomato and the morning glory not all that far from where Baptist Ott formerly grew them. That is not only *in situ* conservation, it is *in family* conservation. At the same time, Kent and his helpers have also grown out an additional two thousand vegetable varieties in the same summer garden, hand-pollinating the outcrossing crops to preserve the genetic purity of each. This may be considered *ex situ* conservation, but with a difference: the Whealys have adopted many of these orphaned varieties, so that they are becoming part of the family's shared experience. The Whealy kids grow up playing with bags and jars of brightly-colored beans, eating delicious watermelons on the porch in the evening, and carving all kinds of pumpkins at Halloween. This diversity is now part of their family legacy. Kent is now hoping to extend his love for heirloom vegetables to other vanishing plants and animals as well. He has initiated an orchard of rare fruit varieties, a herd of White Park cattle, and a prairie restoration project.

If there is any limit to this vision, it is that today, so much of it is dependent upon Kent and just a few others: Diane Whealy, David Cavagnaro, and Glenn Drowns. A dozen other gardeners scattered across the country serve as Kent's Vegetable Curator's Network. Should anything happen to this core of concerned gardeners, how will the Heritage Farm fare next year or in fifty? Will Aaron Whealy or one of his sisters continue on with the Heritage Farm once their parents have passed on?

Although Kent has already saved hundreds of heirlooms from extinction, the outcome of his effort still remains in doubt. Can the Heritage Farm continue to keep these resources, and the folklore surrounding them, alive for many generations to come?

As a complement to Kent's Seed Savers Exchange, four Southwesterners, myself among them, began Native Seeds/SEARCH in the early eighties. We were already well acquainted with a number of farmers in Native American communities in the southwestern United States and northwest Mexico. We realized that these people not only maintained many native crop varieties, but were eager to obtain others that had somehow been lost from their communities. To date, we have helped return seeds to more than a dozen reservations in the Southwest, and to many indigenous communities south of the international boundary as well. In addition, the organization keeps two thousand accessions in medium-term storage, and when possible, surplus seeds of these varieties are sent on to long-term seed storage laboratories and shared with our membership of more than fifteen hundred home gardeners. The more garden plots these seeds are grown in, the higher the probability that they will survive.

But what we are seeing is that the mere reintroduction of seeds to a Native American village does not ensure reestablishment and continued production. The pressures on these communities, particularly in the U.S., are fierce. Economic, political, educational, and psychological forces work against their farming traditions, encouraging young people to migrate to the cities, or to accept government welfare as less demanding alternatives. Many forces are tearing at the fabric of these communities, and yet their renewed interest in self-sufficiency is gaining ground. In the meantime, Native Seeds/ SEARCH and the Seed Savers Exchange serve as bridges, keeping alive seeds of the elders until the younger generation has a chance to prepare the earth for them again.

VI.

I had thought the generation gap between seed savers was endemic only to a developed country like the United States until I learned of a similar project in Tabasco, Mexico. Over the last thirty years, oil drilling, cattle raising, and urbanization have opened up 2.5 million acres in the state of Tabasco, land that was previously covered by rich tropical forest and small farms. The region has been overrun with exotic plants from Asia and Africa, even though it is a center of origin and diversity for many cultivated plants.

José Ramírez-Acosta, an agronomist who has been working in the area for over a decade, laments the genetic erosion that has occurred so swiftly there: "This situation has provoked a marked process of marginalization and in some cases the disappearance of a large set of native genetic resources that Nature has offered to humans in this tropical region—resources that without a doubt represent a source of nutrition for future generations. The garden must assume the functions of a germplasm bank active in conservation, evaluation, and utilization of these vegetatively-propagated resources."

The garden in Tabasco that José speaks of is called the Jardín Agrícola Tropical del CRUSE. It is a tropical version of Kent and Diane Whealy's farm. Beginning with salvaged local varieties in 1974, José has developed a living collection that includes fifty species of native fruit trees, thirty medicinal species, and several root crops. From avocado, ramon, and chocolate through vanilla and pepper, to calanga, macal, and sweet potatoes, José and his coworkers have been stockpiling semi-cultivated food plants and domesticated folk varieties. Although he has not yet received many requests for them from local farmers, José is willing to offer cuttings to any farmer in the region willing to grow them again. His tropical heritage orchard is like a holding tank waiting for the flood of progress to pass in his disrupted region so that he can let the unique, distinct, and pure land races of Tabasco grow back again.

VII.

Will food plants survive such floods? The Kayapo Indians of the Amazon have a tradition of planting a diversity of crops on high ground as "insurance gardens," in case floods should come and wipe out their lowland fields. How long this tradition has been going on is unknown, but ethnobiologist Darrell Posey feels that it arose out of the Kayapo's concern with a mythological flood that potentially threatens to destroy their food supply. To avoid such an event, Kayapo women plant a variety of edible tubers and medicinal plants under bananas in hillside orchard gardens, above the high-water levels reached near their villages.

The Kayapo are a nation of twenty-five hundred Ge-speaking Indians living in nine widely dispersed villages close to the equator. Although they draw upon a considerable variety of wild foods within their five-million-acre indigenous reserve, they still take this extra precaution to preserve their cultivated foods and medicines. The hillside gardens are not carved out of new jungle vegetation. On the contrary, old fields are usually cleared of their underbrush and a representative selection of the village's crop heritage is planted. These heritage gardens are carefully tended by the female elders of each village, under the scrutiny of the highest-ranking woman. Because many of the crops require shade, a canopy of bananas is maintained above them. When a new hillside heritage garden is required, bananas are first transplanted there and then the understory crops are started under these banana trees.

This is conservation as a community responsibility and perhaps also as a sacred duty. These plots are considered vital to the survival of the community that tends them. Other communities and cultures would do well to follow their example.

VIII.

It is a small conceptual leap from the Kayapo gardens to biosphere reserves—a kind of protected area in which traditional uses of plants and animals by indigenous peoples persist without interference from outside pressures. The biosphere reserve concept emerged from UNESCO in the mid-seventies, as a strategy for conserving both natural landscapes where humans have had little impact, and adjacent cultural landscapes where native farmers, hunters, and gatherers have found ways to continue their subsistence activities without depleting scarce resources.

The conventional "national parks" management strategy in the United States seldom allows for traditional land uses in or near the protected area itself. In fact, the National Park Service has historically attempted to buy out Indian enclaves within park boundaries, assuming that their farming, gathering, or hunting activities were "unnatural" or contrary to nature preservation. Some National Park administrators are even reluctant to allow the collection of seeds of known plant genetic resources for backup storage and evaluation in seed banks and botanical gardens, even when there is no danger that this collection will deplete the plant populations.

It comes as no surprise, then, that the biosphere reserve concept has been more immediately embraced in the Third World than in the United States, where the Park Service is attempting to superimpose the concept over existing national parks. In developing countries, where the pristine park concept may be foreign anyway, progress has been made toward combining natural and cultural resource conservation into one scheme. My own experience in Mexico's El Cielo reserve bears this out.

For culturally sensitive nature conservation, the most celebrated example in the Americas today is that of Nusagandi Park, cautiously managed by the Kuna Indians, and proposed as a biosphere reserve. More than thirty-one thousand Kuna speakers live in sixty-five villages scattered along the Caribbean coast of Panama and Colombia. Whereas the riverine Kuna of Colom-

bia have recently lost much of their land to deforestation and modern farming development, the San Blas Kuna of coastal Panama still retain control
over the Comarca de Kuna Yala, an indigenous reserve containing lowland
tropical vegetation and coral islands. However, their indigenous reserve extends up three thousand feet along the Continental Divide, where wet tropical forests cover the mountains. The coastal Kuna do not make intensive use
of large tracts in the mountains, but have nevertheless shown interest in their
preservation for several reasons.

The Nusagandi Park idea emerged from efforts of Kuna youth to stop
encroachment by non-Indians into their reserve. At first, the Kuna attempted to establish an agricultural village as a way of impeding squatters
and forest poachers who might enter their reserve by a new road from Panama City. The site was found to be a difficult one for Kuna subsistence farming, which traditionally involves sixteen crops and forty-eight kinds of useful trees. Reluctant to abandon the site, the Kuna sought other ways to
maintain their land rights. They received support for an indigenously-
managed nature reserve and established contact with scientists who might
assist them in park development. Soon, the beginnings of Nusagandi Park
took root in the virgin forest not far from their village.

One Kuna participant in the park planning has commented that, for the
moment, "the interests of the Kuna and the scientists converge." Kuna activ-
ist Guillermo Archibald believes that his people, like many of the scientists
who have visited them, understand the interdependency between humans
and other organisms: "Kuna shamans use medicinal plants, and teach that
they are living things that think, feel, and hear, friends of the Kuna that exist
for mutual preservation, the death of one spelling the extinction of the
other."

Virgin tropical forests, referred to as *neg serret* in the traditions of the
Kuna, are abodes of powerful spirits and sources of many medicinal plants.
Although small areas are farmed by Kuna families, they have let much of the
vegetation within the Comarca de Kuna Yala reserve remain in a condition
approximating virgin forest. Outside the Comarca, the denuded hillsides of
Panama and adjacent Colombia provide a sad contrast; each year, over one
hundred twenty thousand acres of wet tropical vegetation are being lost in
Panama alone. As a Colombian Kuna leader visiting the Comarca explained

of the trend on unprotected land, "In the beginning there was just virgin forest; but when we looked again, it's not like that anymore. All of the trees have been swept away, and great farms surround us."

To leave extant forest in a wild state is simply not a foreign notion to the Kuna, as it may be to other cultures. In their own cosmology, *Kuna Yala*, 'the People's place,' is an integral part of *Abia Yala*, 'Mother place,' reflecting a dialectic not unlike that of "traditional use area" and "protected core area" of the biosphere reserve model.

Today, about twenty-five people, mostly Kuna, are working on the Project for the Study and Management of Kuna Wildlands, or PEMASKY. Cooperating with internationally funded biologists, the project has identified eighty endangered plants and animals within a few hundred acres of the park. According to a PEMASKY visitor, Barbara Dugelby, fifteen new plant species have been described from collections made by the project's botanists. When visiting botanists study in the reserve, they are accompanied by Kuna guides, and they must turn in reports on the preliminary findings to the Kuna before they leave.

At Nusagandi, 'Place of the Rats,' a few buildings for lodging and research have been constructed, and five modest nature trails have been established. The Kuna assume that the area will soon attract more visiting scientists and occasional tourists. Their efforts will be to orient these visitors toward the rich natural and cultural heritage of the reserve. Species identification labels along the nature trails will give plant names in Latin, Spanish, and Kuna.

It is fitting that the impetus for much of this work came from a young Kuna man, Guillermo Archibald, who, after being trained at an agricultural school in Panama, has returned to his people. It is just as fitting that he is one of the coauthors of a paper in *Cultural Survival Quarterly* about indigenous cultures and protected areas in the Americas. What were once considered separate issues—cultural survival, agricultural stability and diversity, and wildlands preservation—now seem to be tightly intertwined.

Let us keep these three strands wrapped together in a rope that we can climb to rise above the currents of extinction. Let us weave that rope into nets by which we can rescue the cultural, natural, and agricultural resources that are threatened by the floods below.

Part Two

The Local
Parables

CHAPTER SEVEN

Wild-Rice:

The Endangered, the Sacred, and the Tamed

I.

I had driven two hours out of my way to try to see an aquatic grass that seldom, if ever, gets its head above water. Descending out of the Texas Hill Country, I was supposed to be on my way to Houston on the Gulf Coastal Plain. But the story of *Zizania texana*, Texas wild-rice, had pulled me off course and into a landscape too manicured and manipulated for its own good.

As I drove along, idly listening to an Austin country-and-western radio station in my rented car, I hummed a sad song that summed up my pilgrimage. It was "Needle-in-Haystack Time Again," looking for a plant so rare that I'd need a world's expert to take me to it. Fifty years ago, Texas wild-rice was abundant along several miles of the San Marcos River, and in the irrigation ditches flowing from it. A decade ago, Dr. W. H. P. Emery found it only within a thousand square yards of the river—an area about one seventh the size of The Alamo in nearby San Antonio. What's more, streamflow had been increased to the extent that the seedheads, which were formerly raised a yard above the water, are now constantly pummelled by the current so that they remain submerged, incapable of sexual reproduction.

As conservation biologist David Riskind explains it, this Texas wild-rice "is barely hanging on because it is [left with] surviving in a completely alien habitat. It used to thrive as an emergent along a spring run. Since the San Marcos has been dammed, the species must cope with survival as a submerged aquatic." In short, the Texas wild-rice habitat is as converted as a box of Uncle Ben's.

Although this grass is not closely akin to Uncle Ben's white rice—*Oryza sativa*, one of humankind's mainstays—it is related to two minor cultivated crops, *Zizania palustris*, and *Z. latifolia*. Northern wild-rice, *Zizania palustris*, grows around the Great Lakes and surrounding watersheds. Its dark, quill-like grain has been eaten and revered by Algonquian-speaking tribes for millennia. It is now found mixed with true rice in most American markets. Manchurian water-rice, *Z. latifolia*, is grown mostly for its fungus-infected stalks, eaten as a vegetable called *gau-sun*. Its cultivation, however, may originally have been for grain rather than for greens. Manchurian wild-rice is described in Chinese herbals written a thousand years ago. Three thousand years ago, its seeds were featured in rituals of the Chou Dynasty.

As I leaned on the railing of a bridge crossing the San Marcos River, it was evident to me that *Z. texana* had no sacred status locally. The Tonkawa and Comanche bands that once lived along the river are long gone. For the last two-and-a-half centuries, one group of immigrants after another has disrupted the local environment. The so-called "natural range" of this perennial grass is now restricted to the small patches that cover only a thousand square yards along a two mile stretch of the river, entirely within the city limits of San Marcos. And densities have plummeted.

I looked out over a channelized section of the river, where water flowed so swiftly that all aquatic plants seemed plastered under the current. Roundup and other herbicides have been used regularly on the "weeds" around bridges such as the one I stood upon. Chemical pollution, rapid flushes of storm sewer runoff, and mowing of aquatic vegetation have damaged other urban stands of this stream-loving grass.

Worse yet, hydrologists fear that within the next three decades, increased groundwater pumping will dry up the San Marcos Springs in the Edwards aquifer that feeds the river. The accelerated use of the aquifer is an antici-pated result of the rapid population growth within a fifty mile radius around San Marcos. If the San Marcos Springs bite the dust, Texas wild-rice will be one of several rare local species left without a home. An endangered sala-mander, an endangered gambusia fish, and a threatened fountain darter are all endemic to the upper reaches of the San Marcos. The U.S. Fish and Wild-life Service is now attempting to manage the springs and the river as "Critical Habitat" for these four species, but they are dealing with conditions that have been irrevocably disrupted.

A few biologists feel that because they are unable to limit human popu-lation growth within the endangered plant's original habitat, the species should simply be uprooted and moved to a safer place. That is, if a safe site even exists. So far, the track record for Texas wild-rice transplants is rather dismal. A few clumps taken to the USDA's Beltsville Agricultural Research Center failed to last a year and a half in a water tank on the floor of a green-house. Seedlings also died in Maryland, perhaps for lack of proper lighting and water conditions. Closer to home, transplants to Salado Creek in Texas were killed by bulldozers, boaters, and weed removal. Transplants to the nearby Comal River were wiped out by floods. Those placed in Spring Lake were eaten by nutria, an introduced exotic rodent. Where other transplants were made into a selected area on the San Marcos, tubers and canoeists in-advertently knocked over all the wild-rice fruiting heads. All transplants into the wild have been unsuccessful so far. The Endangered Species Recov-ery Team does not now endorse this reintroduction strategy as an effective one for encouraging wild-rice survival.

The only Texas wild-rice transplants that have persisted for any length of time are in a special, water-filled raceway on the campus of Southwest Texas State University. The beautifully manicured campus, home to several biol-ogists concerned with wild-rice, has a walled-in stretch of the San Marcos running right through it. Above the ravaged river, Dr. Emery successfully raised several generations of *Z. texana* in his watered raceway, and has even crossbred some of them with other species of *Zizania*.

Despite the excellence of Dr. Emery's work, it is in some ways a hollow victory. *Zizania texana* is on its way to becoming one of what Robert May and Anne Marie Lyles call "living Latin binomials," captive-bred species that have little left of their former life in the wild. It is a surviving species in name more than in behavior. May and Lyles doubt that such species will ever be able to flourish again if reintroduced into their previous ranges. Even if *Z. texana* habitats were to be more fully protected, it would be difficult to restore them to their original conditions. The full variation in the wild species cannot be genetically reconstituted from the present, diminished natural population, nor from their domestic descendants several generations down the line. A greenhouse tank or campus raceway is radically unlike a spring run of the San Marcos in which this emergent grass once thrived. The wildness has been squeezed out of Texas rice.

II.

"There's someone ricing now," Jim Meeker pointed from the bow of the johnboat. Ahead of us, I could barely distinguish someone in a canoe, pushing into a dense stand of Northern wild-rice. An Ojibway man, pole in hand, sliced into a ripe patch of what he called *manoomin*, the grass known to scientists as *Zizania palustris* variety *palustris*. As the canoe slid into a thinner scatter of rice plants, I noticed his partner sitting in the stern, bending and beating the grass with two cedar sticks.

"Dink. Dink-*dunk*. Dink-*dunk*." We could hear the tap and thud of the ricing sticks as the harvester bent the seedheads with one, then flailed the grain out, rapid-fire, with both. The rhythmic clicking of the sticks against the aluminum canoe resounded across the northern Wisconsin slough. Visually, it was like watching an ancient drum dance, the Ojibway whirling their ricing sticks in tight arcs, keeping a constant sprinkle of seed pattering onto the canoe floor as they moved along.

We pulled up parallel to the harvesters, though we stayed out in the open, faster-moving water. The poler tipped his baseball cap to us, but kept the canoe moving.

Recognizing the men, Jim asked, "Much rice out there to harvest?"

"No, not too much rice left," the rice-knocker replied, as he raked a few more seedheads over the side of the canoe.

"Getting more exercise than rice today?" I asked.

"That's about it!" They laughed. "But it's okay." Their canoe slipped into a thicker stand, and we lost their voices to the push, whisk, and patter of the work.

"Whenever I meet Indians in the slough during the ricing time," Jim said quietly as we pulled away, "there are nothing but smiling faces. They get part of their sustenance out there. It's also a time when they get together. Cousins come in that they haven't seen for a while, and they all get in their canoes."

Jim's words echoed those of an Ojibway man, Ernie Landgren, commenting on wild-rice harvesting over at Nett Lake: "There's a feeling you get out there that's hard to get other places. You're close to Mother Nature, seeing things grow, harvesting the results of the water and sun and winds. . . . We sort of touch our roots when we're among the rice plants."

The wild-rice and Ojibway workers we met that day were rooted in the slow-moving waters and fertile muds of Kakagon Slough. The slough is a broad, low-lying landscape, one in which squawks of herons and quacks of ducks travel far. Kakagon Slough is a national landmark that has been praised in the *Federal Register* as "the finest marsh complex on the Upper Great Lakes." Botanists Jim Meeker and Lisa Ragland have recorded about seven hundred species of plants within the ten vegetation communities in the slough vicinity, where three rivers converge on Lake Superior. Rich in waterfowl, in fish, and in history, Kakagon also offers the Ojibway an abundance of wild-rice, as much as four hundred harvestable acres in any given year.

Kakagon is on the Bad River Indian Reservation, one of eleven Ojibway or Chippewa enclaves in the wild-rice region of Wisconsin. Jim Meeker is investigating wild-rice productivity for this Native American community with logistical support from the Great Lakes Indian Fish and Wildlife Commission. Jim, like the young Ojibway men he works with each summer, is accustomed to *wild* wild-rice: "This isn't a paddy situation; it's naturally sown, in moving waters. Though it occurs in thinner stands, river rice always does well. Little disease ever shows up in these riverside patches." Pathologist James Perchic later confirmed this observation; diseases found in rice paddies are not infecting wild-rice stands in streams less than two hundred yards away.

Meeker, a University of Wisconsin graduate student, has been trying to figure out what factors naturally enhance the reproduction of wild popu-lations relative to others. He also studies how much wild-rice survives from one stage of its life to the next. Today, my wife, Karen, and I have come up from Beartrap Creek with Jim and his neighbor, Greg LeGault. Greg and I had worked together as "environmental planners" fifteen years ago, writing obituaries for marshes on Lake Michigan, before he retreated to the North Woods, which are among his true loves. Now we are helping Jim glean wild-rice seedheads that he had earlier wrapped in cheesecloth. He had protected their developing grains to determine their yields before waterfowl or human harvesters took part away.

"Blackbirds, mallards, sometimes muskrats, affect the rice yield," Jim ex-plained, reaching over the bow to grab a soggy cheesecloth sack filled with seed. A handful of waterfowl species actively consume the grain itself, while dozens of other animals, from snipes to sedge wrens to minks, depend upon it for cover.

Jim pulled the cheesecloth-enclosed seedstalk out of the water, and frowned.

"It worked pretty good, but not good enough. Some of the cheesecloth sacks get buffeted by waves or soaked by rains, and they weigh the seedheads down too much. I hope it doesn't affect my results—I'll have to find a better way to protect the seedheads."

Like most field biologists who have tried to improvise a method for data collection, I knew Jim's frustration. I first thought of the manner in which the Ojibway formerly tied off the seedheads to protect them from birds, leav-ing light strips of basswood bark over them for a couple of weeks while they ripened. Then my mind began to consider ways to cover the seeds using ma-terials that are more readily available today. "Maybe you could use nylon hose instead, and it wouldn't soak up so much water," I suggested.

Greg let out a hoot from his seat at the stern. "I can just see Jim here, going into a store to buy nylon pantyhose for his *research*," he chuckled, "and tell-ing the girl behind the counter that he's gonna put it on some *rice plants*!"

Meeker took the teasing with grace. But he quickly dismissed our com-ments, for his mind had already moved back to the matter at hand. Jim had been sorting out the patterns of rice density and stand size in relation to river meanders and slough sedimentation. He was watching the breadth and length of wild-rice patches as we headed back upstream, from the confluence with Lake Superior, toward the hatchery.

"The best river-rice, I guess, grows where the sediment drops just downstream from the inner curve of the river meander . . ."

Meeker pointed to a twenty-yard-wide stand in slowly moving waters, in the shadow of a bend in Beartrap Creek. Pickerelweed, pondweed, and water lilies were interspersed with the rice, but their densities seemed rather low.

"The broader stands in the slough, or in lakes, are easier harvesting. But here, near the meander, the dynamics of sediment flow allow the rice to outcompete the perennial vegetation."

The sediment influx renews the nutrients that wild-rice depends upon. It's a nitrogen-loving plant, like most annual grasses. But unlike paddy rice, which requires artificial fertilization, the wild-rice in rivers receives a periodic pulse of new nutrients from floods. The sloughs near the river mouth may also benefit from natural perturbations of bottom sediment caused by seiches, which serve to bring water and nutrients in from the confluence with Lake Superior.

According to oral accounts recorded just before the turn of the century, the Indians themselves have long encouraged wild-rice productivity. In 1897, an Ojibway woman, Paskin, told ethnographer A. E. Jenks that the wild-rice stands along the Lac Courte Oreille river and four nearby lakes were originally sown there from a harvest brought in from Prairie Lake in Barron County, Wisconsin. The Lac Courte Oreille community, close kin to those at the Bad River, "have a tradition that all the wild rice between their present habitat and the Red river to the North has been sown by their ancestors," Jenks wrote in his report to the Smithsonian Institution. "The finest harvest field now on the reservation is that of the Lac Courte Oreille river. It is a sown field. . . . Awa'sa sowed the grain [there], and his grandchildren's families now harvest the crop."

This Indian seeding of wild-rice was not an isolated incident in history, according to Jenks. Over a thousand mile stretch from northern Wisconsin west to the other side of Lake Winnipeg in Manitoba, Canada, various tribes intentionally sowed certain marshes. How many times rice was moved to new locations in this way, no one knows. But considering that the most ancient evidence that wild-rice was processed by American cultures is 2500 years old, Woodland tribes had ample time to influence the distribution of this food plant.

Other Ojibway bands may not have influenced the plant's distribution so much, but they have certainly affected its abundance, at least in the short run. At Rice Lake near Crandon, Wisconsin, rice was sometimes sown where another competing grass with large flat stalks was weeded out to ensure better rice harvests. Historian E. J. Danziger, Jr., claims that the Ojibway once "planted about a third of their harvest to ensure a yearly increase."

I asked two young men from Bad River whether intentional resowing of traditional stands is done by the Ojibway today.

"The way we harvest it, much of the seed gets knocked into the water anyway. We don't have to resow it," one explained.

"Not all of it is ripe when we go through in canoe. You'd have to harvest the same stand every three or four days to get it all as it ripened," said the other Ojibway man, "and even then, most of it would end up in the water."

Biological studies confirm these opinions. Traditional hand harvesting skims off only ten to twenty percent of the seeds produced in wild-rice populations. The Great Lakes Indian Fish and Wildlife Commission has concluded that "the conservation practice of traditional rice harvesting therefore leaves ample amounts of rice for reseeding and wildlife utilization."

There is a spiritual dimension to this conservation practice, one which reminds participants of the source of this bounty. At the onset of wild-rice ripening in late summer, a "first fruits" ceremony is offered to thank the *Manitok*, the spirits who inhabit the plants, the water, and the earth. At least in former times, the harvesters on their first outing left a gift of tobacco on the waters, to placate the Water Monster who might otherwise overturn their canoes. Soon after the first gleanings from the rivers and marshes were landed, they were parched, hulled, winnowed, and cooked up for a community feast. In speeches by elders, the harvesters were reminded that the *manoomin* is a gift from the *Manitok*, who urge them to gather it in a "respectful and thrifty manner" and to use it in the spirit in which it is given.

Today, some families may privately make offerings in their own homes, but an intertribal pow-wow hosted by the Ojibway at Odanah carries an element of the ancient ceremony within it. For four nights, a sacred fire is kept burning at the pow-wow, and as one Ojibway keeper-of-the-flame explained, "there, each person can make an offering in his own way."

Some of the same Ojibway men who organize the pow-wow and lead chants for the Bad River Singers also work on the Commission's studies of wild-rice. They will be working to rid the rice beds of purple loosestrife, an exotic weed that has recently been found in fifteen wild-rice populations, and is known to outcompete native marsh plants elsewhere in Wisconsin. In this work, they are protecting an economic resource of their tribe, but they are also preserving their culture. As one of the men unabashedly admitted, "I just love the rhythm of the ricing."

Most of this century, Indians concerned about wild-rice have had to focus on two issues: the diminishing sizes of rice populations, and access to off-reservation populations. In Wisconsin, wild-rice has been so depleted in its natural habitats that the state has declared it a Scarce Resource. In addition to competition by exotic weeds, introduced carp and rusty crayfish damage many populations. The drainage of wetlands, damming of rivers, and chemical pollution—in part from fertilizer overloads in commercial cranberry bogs—have negatively affected wild river-rice. No one hazards a guess as to how much wild-rice has been lost as a result of these disruptions, but few scientists doubt that the reduction in distribution and abundance has been significant.

Another issue—that of Indian rights to off-reservation ricing fields—has recently been the topic of a heated legal battle. In a 1983 decision overturning an earlier opinion by Judge James Doyle, the U.S. Court of Appeals reaffirmed Ojibway usufruct rights to wild-rice fields over a vast area of Wisconsin. Despite their agreement to *live* within certain reservations outlined in treaties in 1837, 1842, and 1854, the Ojibway neither released nor extinguished their interests in gathering this traditional crop throughout their aboriginal territory. After having his earlier decision overturned, Judge Doyle confirmed on February 18, 1987, that this permanent right meant that tribal members in Wisconsin could gather, trade, or sell amounts of wild-rice "to that quantity which will insure them a modest living."

The various Ojibway communities in Wisconsin agreed to manage, conserve, and protect this native resource through developing their own biological expertise, and through using the studies of already established experts in the field. Today, one Bad River resident is applying his college biology degree toward wild-rice management, and others are receiving on-the-job training. The knowledge of competent field ecologists such as Jim Meeker is being accepted as a valuable aid to this cause. It is hoped that the dozens of Ojibway families already accustomed to ricing in Kakagon Slough on the Bad River Reservation will soon feel comfortable with exercising their rights in marshes and rivers elsewhere on public lands.

III.

Ironically, about the time of the decision allowing the Ojibway to earn a modest living off natural stands of wild-rice in northern Wisconsin, the California paddy production of this crop eclipsed the total harvest of all wild stands in the United States and Canada. In 1977, a Minnesota wild-rice variant that retains its seeds longer was introduced into the Sacramento Valley, where Old World domesticated rice had been cultivated since 1910.

By 1983, California cultivation of *Zizania* exceeded a million pounds of marketable product. In 1984, California production more than doubled again, as it did in 1985. In that year, California produced more than half the *Zizania* harvest in North America, surpassing Minnesota in cultivated production for the first time. In 1986, California's fifteen thousand acres of paddies yielded more than ten million pounds of wild-rice, perhaps twenty times the quantity of wild-rice harvested from natural habitats in its native range.

When I first saw bags of wild-rice in roadside fruit markets in the Sacramento Valley a couple of years ago, I did not understand that the grain had been produced locally. Later, when I did grasp the implications of wild-rice cultivation in California, I realized that the entire process of cereal domestication is being telescoped into a few decades with wild-rice. It has jumped from Paleolithic gathering to modern production where engineered irrigation, improved seed, and aerial applications of chemicals have become routine.

Although the Indians have sown and managed wild-rice for centuries, attempts to grow it like other cultivated crops—in fields especially designed for it—did not begin until the fifties.

"It was 1950 when James and Gerald Godward grew a one acre field at Bass Lake near Merrifield, Minnesota," recalls Professor Ervin Oekle, the historian of wild-rice domestication. Others also tried their hands at it, but because of pathogens or other management problems, most of them quickly abandoned the endeavor. "In 1959, the Chun King Corporation near Duluth collected seed from a natural stand and planted a twenty-five acre field which had dikes around it so the field could be flooded. The first two years they obtained good yields but the third year the field was completely destroyed by a disease which turned out to be leaf blight."

Whenever the density and uniformity of a wild plant population is increased, diseases that are seldom found in nature become a threat. But by the sixties, growers had learned to shift their monocultural plots whenever the blight caught up with them, which was usually by the third or fourth year. They were grubstaked by Uncle Ben's of Houston, which had found a favorable market for blends of rice and wild-rice in grocery stores. Commercial cultivation of wild-rice made headway, soon overshadowing the unpredictable returns from wild stands harvested by Indians in Minnesota and Wisconsin.

About the same time, two University of Minnesota agronomists found the first genetic trait in *Zizania* that could move the crop from mere cultivation to true domestication or selection away from the wild type. In the fields of pioneer wild-rice grower Algot Johnson, Minnesota plant scientists found a strain of plants that retained both their male flowers and their seeds longer than the rest. Because "seed shattering" or natural dispersal was reduced in this strain, harvesters could make only one pass through a field and gain a substantial yield. By the time this Johnson variety reached California, where it was further selected for height and late maturation, as much as a thousand pounds per acre could be harvested by mechanical combines. Machine-combining ten acres an hour, a two-man paddy crew can far surpass the 200 to 250 pound harvest of two Ojibway gatherers working all day in a canoe.

There are several reasons why wild rice is yielding so much better in California than in its natural range. First, when an improved variety is sown into a river or marsh in the Great Lakes region, its potential yield is diminished by competition. Where sown in the wild, it exists as a component of a more diverse plant community, and it may hybridize with the unimproved wild-rice there. Californians can completely drain their paddies, chemically kill any weeds, and then reseed and refill the paddies for optimal production. California production expenses are higher per unit area, but not per pound of harvest. A pound of wild-rice has cost Californians as much as two dollars less to grow than it has a Minnesota grower.

Thus rice prices have generally come down as inexpensive California products have hit the market. And yet, both California and Minnesota growers are suffering from the instability of the industry. In California, some growers could not obtain contracts for their production in 1987, while others suffered from crop damage by coots and by new diseases, which together damaged more than two thousand acres of paddies. Inefficient producers, or those who borrowed too much money when rice prices were high, are going bankrupt in both California and the Great Lakes.

"It's gone to hell in a handcart," one *Zizania* specialist said of the market. "I'm afraid to call up a Minnesota grower, for fear that he may not still be in business. . . . And when he goes under, he dumps his remaining rice on the market for next to nothing, and that hurts even the efficient grower who is still in business."

Driving from Duluth to Bad River in 1987, this instability can be seen on handwritten price signs in front of roadside stands. For $1.75 a pound, you can buy cracked paddy rice rejected from Minnesota packagers; perhaps some of it was sent in from California. Black, even-grained Canadian fancy rice may sell for $4.00 a pound. Mixed, Minnesota lake rice is valued at $4.00 a pound.

Then there is the Ojibway man who offers kelly green, mottled, long- and short-grained wild-rice for $6.00 a pound. And his price won't budge.

IV.

Vince Bender and his wife live within walking distance of Beartrap Creek. When he was a child, he would dance inside big iron kettles, wearing clean moccasins, to loosen the hulls from the grain. Today, behind his house, Bender has a small barn full of threshing, sorting, and winnowing machines

that he and his son, a machinist, have either devised or retrofitted for processing wild-rice. Beneath the metal tumbler of his thresher, Vince Bender
keeps a couple of old-time birchbark winnowing baskets on the floor to
catch any spillage. He is uneasy when it comes to explaining his one-of-a-
kind thresher to us. Then he smiles, and skillfully demonstrates how to win-
now the rice from the chaff in one of the birchbark baskets on a still morning.
He explains why winnowing works: "A lot of people say the wind does it,
but that's not true. The weight of the rice does it." As do the perfectly timed
hand movements of the craftsman shaking the basket.

His wife later explains that Vince Bender is a precise man who clocks the
threshing and parching of the rice to the minute. He knows how much he
can process in a day, how much of the profit should go to the harvesters, and
how much his family deserves. His price, at the back door, is $6.00 a pound
this summer.

When I mentioned the lower prices down the road, he dismissed them as
irrelevant, for those markets aren't selling the same thing. "It tastes like
muck, that rice. It may look good to you—even and blackened. I don't even
know if they roast it. . . . But ours is mixed and green and broken. It tastes
real good."

Mrs. Bender explained in more detail. "Wild-rice needs the running
water to have flavor. Paddy rice doesn't even taste. It's chemically produced."
Visiting an out-of-state processing plant once, the Benders saw "the rice
molding in a big pile, steam coming up from it. We wouldn't want to sell that
to our customers. They leave it in the bag too long and it all turns black. We
like the green color and the taste that you get when it comes from a stream."

These are not merely rehearsed retorts to degrade their competition, but
common reactions among the Ojibway, even those whose livelihoods are not
dependent upon the harvest. One Bad River resident told me that "The first
time I tasted [cultivated] lake rice, I wasn't sure it was the same food. I wondered if the ladies preparing it didn't know how to cook it right."

Marketing propaganda from the Great Lakes Indian Fish and Wildlife
Commission does, however, capitalize on these perceptions: "Compare the
quality of real wild rice with paddy-grown 'wild' rice. Most of the wild rice
on the market . . . is actually cultivated . . . using fertilizers, herbicides, and

machine processing to maintain high production levels. You can recognize it by its blackened kernels, and muddy or gamey flavor. While more expensive, Lake Superior Gourmet Wild Rice is grown organically and is traditionally harvested and processed by ancient Chippewa methods. Observe its lighter grains and more delicate aroma and flavor . . ."

At White Earth, Minnesota, Winona LaDuke of the Ikwe Marketing Collective for the Anishinabey Indians is blunt about the differences: "Most commercial 'wild-rice' isn't wild rice, it's 'tame rice,' a hybrid version of *Zizania*. . . . Tame rice is grown in a diked paddy, with nitrogen fertilizers, fungicides, 2-4-D (a herbicide), and harvested with half-million-dollar combines."

Despite the contrasts in production and marketing, the California and Native American products have many of the same benefits and costs in the eyes of the average American consumer. For certain nutrients, cultivated *Zizania* seeds are not measurably different from the wild seeds, but both are richer in protein and lysine than most commercial cultivars of corn, brown rice, rye, and barley. All wild-rice costs considerably more than these other cereals, for it is regarded as a specialty food, not a staple, even among most Indian families.

At another level, however, the difference is one of deep-seated values, not merely economics or narrowly defined consumer preferences. Poorer Indian families who cannot afford to buy other specialty items at high prices will spend days slowly harvesting and processing wild river rice—only to give most of it away to friends and relations. Winona LaDuke of the Anishinabey explains that "by consuming our resources, we get a 'use value.' This value—whether from eating wild-rice or berries—is critical to a poor community. . . . To feed our families, we might as well eat good Native food, instead of trying to get the money to buy 'White' food." For many Indians, it is not merely the absence of Malathion or Roundup in the rice beds that matters as much as it is knowing that their sustenance comes from the sloughs and marshes. These wild places are the touchstones of their culture, providing a sense of belonging to Nature that is not provided by a California paddy or a raceway on the San Marcos.

"Nothing can equal the aroma of a ricing camp," wrote an Ojibway woman, Lolita Taylor, a decade ago. She tells us why, not with arguments, but with images: "Wood fires burning, rice drying, and the dewy fresh air drifting in from the lake. A contented feeling of well-being filled the camp. The first grain of the season had been offered for a blessing from the Great Spirit. The time had come to partake of the gift. Boiled with venison or with ducks or rice hens, it was nourishing and delicious."

CHAPTER EIGHT

The Exile and the Holy Anomaly:

Wild American Sunflowers

I.

Most people go to Napa County, California, for the wine, not for the oil that can be pressed from one of the world's more obscure wild sunflowers. Going north from the fertile viticultural valleys, the wine-tasting traffic gradually thins out as narrow roads wind into the Inner Coast Range. I was surprised to find it so unpopulated and so little of it cultivated. It struck me that northern Napa County's dwarf chaparral had its own backwoods charm. I sensed that this charm is geological as much as it is botanical.

Because of the glistening, soapy serpentine rocks that form chunks of this range, plants face mineral problems here—too much magnesium relative to calcium, too much nickel, chromium, and cobalt. This serpentine stretch of coastal mountain range is relatively stark, with stunted shrubs and exposed, thin soils. The trees and shrubs that take to these slopes are a peculiar mix: the leather oak whose leaves curl under at the edges, the gangly Digger pine,

Sargent cypress, chamise, and a rather localized manzanita. Many of the plants are "edaphic endemics," or specialists that have come upon ways to deal with the problems peculiar to being rooted in a serpentine setting.

While serpentine soils and outcrops cover less than one percent of California, geobotanist Arthur Kruckeberg has counted over 180 kinds of plants supported solely by these substrates, notwithstanding the difficulty of their getting a balanced ration of minerals. It is this kind of stimulus that drives the divergence of plants into new forms, which, when isolated in time and space, develop into distinct species. But alas, as Kruckeberg vigorously reminds us, serpentine also attracts miners of gold, quicksilver, nickel, asbestos, and magnesite. Mining operations may have a disastrous impact on rare plants, because these plants target such a limited environmental type as their turf.

Within one such mining zone along the Inner Coast Range, I found wild annual sunflowers growing below remnant serpentine outcrops and on the edges of eroded, mineral-stained gullies. As I looked closely at these sunflowers, I could see features distinct from those of their more common roadside relatives. I knew in advance that if I found sunflowers with narrow leaves and bracts, plus tiny achenes on smallish flowers with few petal-like rays, I could be sure I had stumbled upon a special serpentine form. While a few botanists in the past have considered these serpentine-loving sunflowers to be merely stunted variants of a common California weed, *Helianthus bolanderi*, Loren Rieseberg has recently estimated from chloroplast DNA studies that the weed may have diverged from the serpentine sunflowers as much as three million years ago. Because it has so many unique genetic characteristics, the serpentine-adapted form is considered by many botanists and geneticists to be a fully distinct species, *Helianthus exilis*.

Exiles they are, for gold mining operations have already extirpated thirteen of the sixty-six sunflower populations known previously in this area. Seeds were collected from at least four of these populations before recent land clearing. The mining company itself has helped to fence two of the remaining sunflower stands and has agreed to avoid damaging certain additional places where this plant and other serpentine endemics grow.

Despite the intensity of mining activity, there may be no immediate danger that the exiled sunflower species will become extinct. Though the species may survive, some biologists nevertheless lament the loss of several of its populations. Rick Kesseli of the University of California at Davis devoted a good part of a year to studying the genetic variation of seed populations taken near the mines. "*Helianthus exilis* is very variable, both within and between populations," Kesseli said. He added that "some populations of it have unique alleles," referring to the alternate forms of genes that are responsible for the passing on of particular traits to a plant's progeny.

Kesseli's mentor, world-famous geneticist Subodh Jain, also underscores the importance of each exiled sunflower population. Dr. Jain's decade-long interest in this species is due in part to the unusually valuable genetic traits found within the populations in the vicinity of the mines.

"There is unique fatty acid stability in *exilis* that we can transfer into cultivated sunflowers to improve their oilseed quality," Jain says. Whereas the percentage of linoleic acid in the seed of most sunflowers varies greatly in response to night temperatures during growth, the percentage in the seed of *exilis* hardly varies. And in 1977, Jain and his students also discovered that *exilis* seeds contain a high percentage of linoleic acid in exiled sunflower's seed. Since then, more than twenty other kinds of wild sunflowers have been screened for this characteristic. None has surpassed *exilis* in linoleic fatty acid scores.

At first glance, this may seem to be a fact of interest only to a biochemist with a penchant for greasy subjects. But a superior linoleic content gives a vegetable oil a high ratio of polyunsaturated to saturated fats. Diets in which polyunsaturated oils prevail over saturated fats with high cholesterol may prevent heart disease, the greatest single cause of death in Western societies. Since the early 1970s, crop geneticists have been attempting to breed edible oilseeds higher in polyunsaturates in order to reduce atherosclerosis and coronary complications in consumers.

Of all the genetic materials now available to sunflower breeders that could decrease saturated fats in their products, the exiled sunflower is the most outstanding. A former student member of Jain's team, A. M. Olivieri, has recently accomplished the necessary first steps for exiled sunflower gene transfer, using seedstocks he took with him to Padua, Italy. There, he crossed the exiled sunflower with the closely related but distinct species, *Helianthus bolanderi*, a plant widespread in California.

This hybrid shows a few abnormalities, further indicating some evolutionary distance between the two California sunflowers. Nevertheless, fertile seeds were obtained. Olivieri then used progeny from this cross in hand pollinations of cultivated sunflowers, and must now work with their hybrid offspring to assure that the fatty acid stability and high linoleic content have been transferred. Because oil-type sunflowers are grown on over two million acres in the United States today, Olivieri's work will be of great value to farmers as soon as the exiled sunflower genes are fitted into the right commercial cultivar.

II.

As California gold miners drive home from work each day, they pass the remaining stands of the exiled sunflower. Charlie Rogers, a specialist on sunflowers and their associated insects, recalled to me the time he was collecting seeds of *exilis* along the roadside not far from the mine. A truck drove up and a miner stuck his head out of the window.

"Hey man, tell me where I can get a license."

"A license for what?" asked Charlie, continuing his business.

"Are you high or something? *Doves!* A hunting license. It's dove season! What are you doing there?"

"I'm collecting sunflower seeds," Charlie replied, still moving along through the stand.

"I can't believe my ears! It's dove season, and this guy is hunting sunflowers. You must be high *and* out of it!" The driver roared off.

If you told these gold miners that the genes from these plants might be more valuable than the precious metal in the ground beneath them, few would immediately believe you. The limp golden rays of the exiled sunflower simply do not provoke the same response in a man as a handful of gold ore. Not many people realize that the sunflower crop, highly dependent upon genes from wild species native to North America, is valued at more than a billion dollars annually.

Nevertheless, the justifications for conserving these sunflowers are not all economic. Joe Callizo, a Napa Valley nurseryman and amateur botanist, loves them because they are unpredictable waifs.

"I have a hunch," Joe muses, "that their seeds are very long-lived in the soil." The exiled sunflowers surprised him one year when they appeared in a spot recently bladed for pipeline construction. Robust, multiheaded annual sunflowers suddenly covered the wounded earth, adding new hues to a landscape long familiar to Joe. "I had been by there all my life. . . . It hadn't been growing all those years, or I would have known it. The color was outstanding."

As a teenager, Joe was impressed by a neighboring farmer's curiosity about the wild plants that volunteered in his fields. Joe recalls how this farmer "would even stop the harvester to climb down to see some odd plant that happened to catch his eye. Later, in college, I learned that the plants which I had seen this farmer examine all had names. What a revelation!"

It was years after his college days that Joe first put name and plant together for *Helianthus exilis*. Since 1983, Joe and his friend Glenn Clifton have been monitoring the status of the exiled sunflower and twenty-five other species of rare plants in close proximity to the gold mining activity. Much of this work has been done voluntarily through a chapter of the California Native Plant Society, which Joe has led. Fearful that the original mining plans would have extirpated over 125 of the 400 populations of the rare plants endemic to serpentine soils of the area, Joe and Glenn offered suggestions for reducing the number destroyed. They tagged plants and flagged the perimeters of populations so that bulldozers could avoid rolling over them when other routes were available. Twice a year, they revisit the sites to assess impending dangers from mining, related erosion, or other factors.

"We have played a chess game with the mining company," Joe admits, as he recalls how certain rare plant stands had to be sacrificed to allow others to be spared. Serpentine seeps where seven or eight rare species grow together have been fenced, and thirty-nine sites continue to be monitored. They have saved nearly fifty populations that would have been obliterated by the original mining scheme. Several of the sunflower stands have survived thanks to their good work.

Joe and Glenn believe that the twenty-five other rare species have just as much right to exist as the more economically valuable *Helianthus exilis*. Their protection efforts do not favor the sunflower over the others, some of which are threatened to a much greater degree. But until the time comes when the general public shares the concerns of Joe Callizo and Glenn Clifton, the potential economic value of this oilseed resource may be a way of deflating one-sided arguments in favor of gold over plants. According to Christine and Robert Prescott-Allen, gene transfer from other wild sunflower species has already resulted in an 88-million-dollar-a-year contribution to the U.S. sunflower industry. There is reason to believe that the work begun with *exilis* by Olivieri and Jain will result in significant benefits in due time.

If such arguments can provide additional incentives for protecting exiled sunflower populations, other rare plants on the same serpentine soils will be saved too. As long as people remain in power who are swayed only by economic arguments, Joe and Glenn cannot afford to dispense with the parable of the sunflower.

Nevertheless, *Helianthus exilis* and other wild sunflowers are not merely economic resources; they are also lifeforms with an intrinsic right to exist. Two wildlife conservationists, Robert and Christine Prescott-Allen, have argued that there is no need to deny either the intrinsic or the extrinsic (economic) values of a species. In their book *The First Resource*, they note that "Many aboriginal peoples live with this dichotomy well enough, expressing strong spiritual bonds with the species that feed, clothe, and cure them as well as with those that make no identifiable contribution to their lives. Improving our understanding of how people benefit from wild plants and animals should not impede the development of a sense of companionship with them. Rather, it should increase our sense of obligation to them."

Where the winding roads of the Inner Coast Range weave past a sprinkle of golden-rayed sunflowers amid the tough-leaved chaparral, I hope that the human sense of obligation will grow. If it does, the exiled sunflower may return to survive and thrive in its homeland once more.

III.

If pressed to name where such a sense of cultural responsibility for wild sunflowers is already in place, I can only reply, "Qa'qa wungu."

Below one of the Hopi Mesas, there's a valley called that—in English, "Place of Many Sunflowers." The sunflower referred to has seeds and heads unusually large for a wild species, hence the name *Helianthus anomalus*. Documented as growing around the Mesas for no less than a century, it is known from fewer than twenty-five Utah and Arizona locations, two of them in this valley.

The anomalous sunflower also grows above, on the mesa, in a special place—around the lip of a kiva, the underground ceremonial chamber of the Hopi. Whether it was planted there on purpose, or sprouted there by chance, this rare sunflower has persisted in the midst of the Hopi fields and villages, despite all that has happened with the passage of time. The Hopi relationship with sunflowers here is not merely a passing association, since certain of these Hopi pueblos are among the oldest continuously inhabited villages in North America.

Within these villages, Hopi maidens customarily dance in the Lagon and Oaqol ceremonies during October and November. To prepare for these dances, they gather wild sunflower petals, dry, and then grind them into a yellow powder. They then wet their faces, and the powder is applied to their skin prior to dressing in elaborate costumes for the dance. Their faces glisten like gold. This Hopi gold is gleaned from the anomalous sunflower.

I first saw this sunflower while collecting Indian strains of cultivated sunflowers for the U.S. Department of Agriculture. I had been given a gift of a few seeds of the Hopi blue dye sunflower earlier in the day at one of the villages. Winding down the road on the side of one of the mesas, I decided to pull off and look at the fields below. But between them and where I parked my pickup, I noticed some wild sunflowers growing in a blowout in the sand dunes. I scrambled down to them, picked off a small side branch, and placed it in my plant press. Assuming it was the common weedy *Helianthus annuus*, I forgot about it for nearly a year. I was supposed to be collecting seeds of Indian crops, not weeds, although I would occasionally throw a few weedy associates of crops into my press. At that time, despite its affinity to the domesticated sunflowers that I was after, I did not assign any more significance to this wildling than I would have to any other.

During the following year, I received a message from the team of sunflower researchers based at Bushland, Texas. They had been successful in obtaining seed of most sunflower species on the continent, but had unfortunately failed to collect seed of *Helianthus anomalus* in the vicinity of the

Hopi Indian Reservation. Charlie Rogers wrote me that he had seen "only a single glistening plant from the highway below the big mesas." Aware that I had collected domesticated sunflowers there, he asked if I was willing to go up and search for seeds of this rare wild species.

Before I followed through on the request, I checked the herbarium specimen I had made the season before and discovered that I had collected the anomalous sunflower without realizing it. Remembering the location, I agreed to return for seed.

My next trip up to the mesas was during the fall harvest time. Fortunately, a Hopi farmer happened to be out in the field closest to the few sunflowers I had seen the year before. As I walked toward him to ask if it was all right to collect some seed there, I noticed that there were even more wild sunflowers growing on the field edges.

I explained what I had hoped to do, and asked who worked the fields. The man explained that most of the sunflowers were on the edge of his cousin's fields this year, but that it seemed all right for me to take a few seeds from each of the plants around there. He recalled that they formerly grew above the fields, on the snakeweed-covered slopes of the mesa. There, the sunflowers had been so abundant when he was young that he and other Hopi children would regularly scramble through the "tunnels" between their overhanging branches. Whenever they came up around his field, he let them be.

This was curious to me, since his annual crops and his orchard in the sand dunes were relatively weed-free. "Why do you hoe out most weeds, but not this one?" I asked him.

"The place is really named after those sunflowers, so that's why I leave them there," the Hopi farmer answered.

"I'm glad you do," I replied. "There are people in other parts of the world who are interested in that sunflower."

Perhaps because he sometimes grew domesticated dye sunflowers, or because he was generally interested in plants, the farmer tolerated my seed collection and measurement of various sunflower volunteers in his fields. Over the following four years, we would see each other seasonally, as I came up to the mesas to bring native crop seeds to friends, and to keep an eye on this and other sunflower stands. Whenever we sat down for coffee, piki, or a slice of watermelon, the farmer and I would talk of many things: his katsina carvings, our gardens, old paperback novels, and travels. And when the wild sunflower cropped up in our conversations, I would always try to express my gratitude not only for the chance to see it, but for its continued presence.

Over those years, I watched the number of anomalous sunflowers in his field grow from eighteen to over three hundred individual plants. He would leave patches of them out among his crops if they happened to sprout there, while others volunteered on the field margins, or between his peach trees and grapevine hummocks.

These plants tended to be taller and to have more flowerheads than ones from specimens collected earlier in the century from northern Arizona and adjacent Utah. Curiously, the anomalous sunflowers from Hopi fields also have longer ray flowers. These are the golden "petals" that were used in Hopi ceremonies.

Long petals won't make them any more valuable to plant breeders, but their chemical and physiological qualities may be important. *Helianthus anomalus* appears to be a good source of genes for tolerance to drought, heat, and alkalinity, for resistance to lodging, and for high polyunsaturated oil quality. Its potential has not, as yet, been investigated as much as that of other wild sunflowers. However rare, this species is clearly not as threatened as other wild sunflower taxa under observation by the Office of Endangered Species. The protection afforded to the anomalous sunflower by several Hopi farmers is possibly enough to keep it from being officially listed as endangered—even if other populations were to be devastated by livestock, mining, or land clearing. However, if Hopi agriculture ever switched over to pre-emergent herbicides and frequent, deep plowing with tractors, this plant could easily disappear.

There may be no immediate need to worry about the rare anomalous sunflower, thanks to the continuity between generations of Hopi farmers. Still, a dozen other rare taxa urgently require our concern. They constitute nearly a quarter of the species in the genus *Helianthus* in the United States. One distinctive subspecies, the Los Angeles sunflower, already has been extirpated by suburban sprawl and marsh draining. Another species, from the dry lands of New Mexico, has not been relocated for decades, and is presumed to be extinct. The remaining rare species receive little protection in habitat; at best, a few seeds are stuffed into an envelope and considered "saved."

I felt sad when I first realized that so many wild species are threatened. They are so often used as *the* classic examples of wild plants that contribute to crop improvement. To date, wild *Helianthus* species have been used for powdery mildew and rust resistance, for recessive branching, and for cytoplasmic male sterility, a trait necessary for producing hybrid sunflowers. Resistance to rots, wilts, broomrapes, aphids, sunflower moths, and leafhoppers has been found in wild *Helianthus*, and is being transferred to cultivated sunflowers or to their relatives, the Jerusalem artichokes.

Nonetheless, a number of the rarer species have never been screened for any of their useful features, because of the difficulty of obtaining their seldom collected seeds. And for the species that have been collected for frozen seed storage banks, the populations that remain in the wild may be destroyed without regard to their value. However, it is true that the U.S. Department of Agriculture's collection of wild sunflower species is its best effort so far for *ex situ* conservation of a suite of genetic resources native to this continent. The National Plant Genetic Resources Board advisory to the USDA claims that high-tech gene banks are entirely adequate for conserving wild sunflowers and other crop relatives. For some reason, it has shied away from advocating *in situ* conservation of plant genetic resources, even though the USDA administers many lands that harbor species needed by breeders. USDA publications even claim that *in situ* conservation is too unstable and lacking in long-term continuity to be a worthwhile investment.

However risky *in situ* conservation may appear in a time of rapid acculturation and economic upheaval, the USDA could learn a bit about stability from Joe Callizo or a Hopi farmer. Much of the USDA wild sunflower collection assembled at Bushland, Texas, is now being moved, since several of its key caretakers have been reassigned to other research areas. Another wild sunflower project, based for years at the University of California, has been largely dismantled although part of its work continues at Fargo, North Dakota. The brightest students from that program were retrained to work with rice and lettuce when sunflower money dried up on the Davis campus. They cannot obtain jobs with commercial sunflower-breeding firms, for these shy away from long-term, slow-payoff projects. USDA-funded wild sunflower work has been gutted to the point that it can hardly tout its own stability.

Despite the changes that have occurred in their universities and USDA laboratories, sunflower researchers such as Charlie Heiser, Charlie Rogers, Gerald Seiler, John Chandler, Tommy Thompson, and C. C. Jan have given conservationists many reasons to work for better protection of sunflowers and their habitats. Even though some of these men were "reassigned" or forced to move on from their *Helianthus* studies, they have still retained a wonder over sunflowers, in all their diverse forms. If allowed to pursue this fascination in one place for more time, they would be the kind of men who would further the conservation of, as well as the knowledge about, particular sunflowers.

IV.

On a sunny October day in Fargo, North Dakota, I had the pleasure of wandering through C. C. Jan's greenhouses of perennial and annual sunflowers. Whatever awareness I have had of sunflower diversity beforehand was overwhelmed and humbled within a minute of putting my foot in the door. I stood among nine-foot-tall monsters that had not yet reached their flowering stage, and foot-tall dwarfs already spent. Some species had flower heads the size of the palm of my hand, others, heads the size of my thumbnail. Leaf sizes and shapes ranged from arrow-shaped miniatures to wide, fan-like giants.

I stood there for a few minutes, awed, hidden between the rows of perennial species, breathing in warm, humid air with them. Hidden from me that moment were the landscapes that shaped this diversity of forms—the serpentine outcrops of the Inner Coast Range, the windswept Hopi Mesas, the alkaline floodplain of the Rio Pecos, the shores of the Florida Keys, and sand dunes surrounding the Colorado River delta. I closed my eyes for a moment, trying to root myself within those habitats. The sun above warmed me as I faced it. I opened my eyes again, and the world flashed green.

CHAPTER NINE

Lost Gourds and Spent Soils on the Shores of Okeechobee

I.

I pulled the car over just past a road-killed possum, under a sixty-foot arching trunk of live oak veiled in Spanish Moss. I had stopped near the place where naturalists John and William Bartram signed a treaty with the Creek Indians on November 18, 1765, thereby allowing the father and son to continue southward on their Florida Plant Survey. Walking across the country highway, I stared saucer-eyed at the St. John's River, which seemed a mile wide to a desert dweller like myself. Here, the division between river and forest was abrupt. On one side of me was open water; on the other, there was the shade of pines mixed with an assortment of hardwoods, all underlain by saw palmettos.

Nearby, however, the land and the river graded into tannin-stained sloughs. These backwater habitats reminded me of those described in William Bartram's journal. In 1774, the junior Bartram described an area upstream from where I stood, closer to Lake Dexter:

> After going 4 or 5 miles, the land still Swamp and Marshes, observed abundance of Alegators almost every where bask in the sun on the banks. . . . Came to a high Bluff; here the main Land on the west side came to the River . . . went a shore & ascended the Hill . . . left Palm hill & continued up the River, passing by Swamps & Marshes on each side . . . observed the Trees along the River Banks adorned with garlands of various species of Convolvulus, Ipomea, Eupatorium scandens, and a Species of Cucurbita which ran & spread over bushes & Trees 20 or 30 yards high, altogether affording a varied Novel scene exhibiting Natural Vistas, Labyrinths and Alcoves varied with fine flowering plants . . . all which reflect on the still surface of the River a very rich and Gay picture.

I watched ospreys flying out over the river and spotted a reclusive heron partially hidden in a small clump of cattails. Was the scene I saw today as rich and varied as that which the Bartrams saw two centuries ago? I knew that at least one element of the St. John's River flora was now missing; in fact, it was altogether rare in modern-day Florida. The garland had lost its species of *Cucurbita*, a vine that William Bartram described as "the wild squash climbing over the lofty limbs of the trees; their yellow fruit somewhat of the size and figure of a large orange."

This wild squash could have been the feral gourd, *Cucurbita pepo* variety *texana*. Although absent in Florida's wild vegetation today, it was recently identified from archaeological remains at Hontoon Island in Lake Dexter, near the St. John's headwaters. On the other hand, several Florida botanists have suggested that Bartram may have seen a different, more enigmatic wild gourd. These botanists suggest that Bartram was actually referring to a hard-shelled gourd later described as *Cucurbita okeechobeensis*, a tropical plant that reaches its northern limits in Florida. It is not closely related to the widespread pumpkins and gourds of the species *Cucurbita pepo*, such as those found prehistorically on Hontoon Island. It is more akin to a wild gourd of southeastern Mexico, first described as *Cucurbita martinezii*. Recently, geneticist Tom Andres and I have determined that the Mexican *martinezii* gourds and the Floridian *okeechobeensis* gourds should be considered distinctive subspecies of the same species, *Cucurbita okeechobeensis*.

These two subspecies share many characteristics; in fact, they are the only wild American gourds with cream-colored flowers. They are cross-compatible, but exhibit certain differences in their enzymes, in amounts of intensely bitter chemicals known as cucurbitacins, and in the oil content of their seeds. In addition, they have been geographically isolated from one another for centuries. While subspecies *martinezii* is locally common in habitats in eastern Mexico that have long been managed or disturbed by humans, subspecies *okeechobeensis* is rare, and has never been collected from more than a handful of localities. Oddly, these localities are places where gourds have had a long association with humankind.

That history of association has been obscured, however, as Florida landscapes have become drastically altered by modern enterprise. Although I was eager to learn more of the historical context of these gourds, I understood the old adage that "a lot of water had already passed under the bridge" as far as Florida's native cultures and plants were concerned. If I were to find any gourds at all in Florida, I would have to search the tropical shores of Lake Okeechobee.

As I returned to the car and began my journey to one of the largest lakes in the southern United States, I realized that I might not find on Okeechobee the shady tranquility of the forests along the St. John's shoreline. I would discover, two days later, that my premonition was correct. Lake Okeechobee was a combat zone, where environmental degradation and cultural disruption occur at rates seldom experienced in other parts of the United States. Although the Okeechobee gourd was still considered part of the flora of the lake's perimeter, its presence had not been reported in six years. Viable seeds of the gourd itself had not been collected by botanists for a decade. Had this little gourd become a casualty on the battlefield between man and wilderness? I went to find out.

II.

After coming through the Brighton Reservation of the Seminoles, then crossing Fish-Eating Creek, I felt at first as though I were entering another country. The levees and dikes loomed large, while canals and drainage ditches seemed to flow in every direction. The Lake Okeechobee region is just as historian Nelson Blake has described it, "an extreme example of . . . human command over nature. Although this body of water covers almost 700 square miles . . . it is almost impossible to get a glimpse of it from nearby highways. Surrounded by a huge wall of earth and masonry, the vast expanse of water can be viewed only by driving to the top of this barrier."

My guide, Jono Miller, had spent fifteen years investigating the ecological changes in wetlands and islands around the Florida peninsula. I first met him in 1973 when we were both students of island biogeography in the Galapagos. Since then, he and his wife, Julie Morris, had become coordinators of the Environmental Studies Program at New College of the University of South Florida. For our trip, he was furnishing his canoe and pickup, his fine sense of humor, and a long-held curiosity about gourds. For my part, I brought beer, hummus, hot sauce, notes with a few vague directions to where gourds had been seen in the past, and the inherent clumsiness of someone who had not been around a lot of water for quite a while.

After choosing a shoreline campground for our point of departure the following morning, Jono offered the marina attendant there a "wanted" poster he had made for *Cucurbita okeechobeensis*. The yellow flier pictured the Okeechobee gourd, asked for "leads" on its whereabouts, and requested that instead of disturbing the plant, those familiar with it should call Jono's answering service. But before we could get it fastened to the bulletin board, the picture elicited a response from the lady at the desk.

"Oh, people bring those things in here now 'n then. They must grow right around here some place! Let's see now . . . who could you talk to?"

We were finally introduced to Margaret LePelley, who had been coming to the lake for years.

"That plant . . . oh, so it's a gourd! It's on a vine, a very tough vine. I know, because I tried to pick one when we were out boating one time along the islands. I saw this thing hanging out over the water. I tried to reach it by leaning out over the edge of the boat so's I could pull it off. I thought it would come off like an orange—you know, just like that—but it stayed on the vine and pulled me over the edge into the water. I'll never forget that!"

So here were a couple of elderly women casually talking about a plant that botanists hadn't seen in six years, one that had been proposed (and post-poned) for listing as a federally endangered species the Thanksgiving be-fore! "Is the plant locally common?" we asked.

"Well, it's been a couple of years since I seen it, and then it was just here and there. But where it was, it would kind of take over and crawl on every-thing. It would be along the shore where there would be dry ground. I sup-pose it would always be there, but for the fact that the water rises and lowers. Seems like it's been high ever since last summer. Well, I do hope you boys find something!"

Those elderly women with their homespun candor convinced us that we had a chance of finding the gourd, a hope that a dozen calls to botanists had failed to convey. We put up the wanted poster, and did some "wilderness camping," as they call it at the marina, on mowed grass beneath a streetlight, with airplanes overhead spraying insecticides for mosquitoes. Our tent's lack of hookups for electricity, water, and sewer qualified us as wilderness trail-blazers in a jungle of travel trailers, motorhomes, and vans. The concrete lid of the RV's pumpout station made a level, if somewhat fragrant, table for our night's repast. That night, I dreamed of squashes as big as Winnebagos, dying of powdery mildew, while tiny wild gourds survived nearby.

III.

A coot chuckled, a tree duck whistled, a limpkin cried. Jono repeated back the sounds unconsciously as he paddled along in the stern of the canoe, eye-ing the nearest strands of vegetation. While still a boy, Jono had learned the sights and sounds of waterfowl under his father's guidance. Equipped with an anachronistic long-billed swordfishing cap, Jono guided me out among his web-footed friends. It was hard for me to keep my mind on gourds and paddles with all the birdlife around us: rails, purple gallinules, anhingas, glossy ibises, egrets, wood storks, and several herons.

The wind kicked up white caps and made for slow going on open water. Canoeing took less effort once we found a few narrow channels between dikes and island shoals, and our botanizing began. Water lettuce, cattails, hy-acinths, waterlilies, and pennyworts edged the canals. Where Everglades-style airboats had blasted through this floating cover, narrow paths could be navigated by boats with small outboards, or by our motorless canoe. On the south side of the lake, aquatic plants rimmed both the shoreline and four sizable islands nearby.

Even the South Florida Water Management District's bureaucrats have recognized the plants of this littoral zone as "the lifeblood of the lake." During two thirds of its 6300 year history, Lake Okeechobee has provided nutrient supplies and depth profiles adequate to support sizable masses of aquatic vegetation. These plants, in turn, support both the fisheries and waterfowl. The sawgrasses, the floating islands of cattails and willows, and the diverse land plants along the channel banks, all have enriched the organic soils over the centuries. When Jono and I canoed up through a cut in a levee, we could see more than four feet of deep, black, friable earth.

"It looks like the milorganite they sell you in bags, doesn't it?" Jono commented.

Just then, we came upon a fisherman. He offered us the Floridian phrase that has virtually replaced "hello" in the local vocabulary.

"Catch anything?"

"No, we're rotten fishermen," I said apologetically. "We didn't even bring our poles. We're looking for plants."

"Well," the man laughed heartily, "If all of the fishermen who were unlucky resorted to plant hunting, there wouldn't be any plants left in the whole world! What kind of plants are you aiming to catch?"

Jono handed me a poster to pass from canoe to motorized skiff. "It's a little gourd that hangs on vines covering trees."

"Well, now," he said, uncapping his head and scratching his thinning hair. "That may be what my daughter has wanted me to get her for making into some kind of decoration. She keeps on describing this gourd, and I keep on bringing her those lotus seedstalks, and she keeps on saying that they aren't what she wants. Maybe this is what she's talking about!"

Women, we discovered, have been the ones who have noticed the gourds in the past. Most men are singularly preoccupied by fish. As we pulled away, this man was still scratching his head, looking at the poster, his line bobbing without any indication that even one fish cared in the least about it. We drifted down a channel along a high dike that once encircled the island, keeping it dry enough for settlers to grow corn there. Though both the farmers and the corn are now gone, other plants that the farmers brought with them remain: castor beans, papayas, bananas, casaurinas, and sow thistles.

We passed another steep-sided dike, and I thought about the amount of buried lake bottom soil that had been dredged up and placed in contact with air. That exposed dike was a microcosm of the entire Kissimee River–Lake Okeechobee–Everglades watershed, for the history there has been one of separating water from land. European-Americans decided a century ago that the lake's periphery offered the largest reclaimable mass of organic soils left unexploited on the continent. Since then, reclamation projects have drained thousands of square miles adjacent to the lake and channelized rivers into straight canals. These projects have shaped the entire watershed from Disney World to Everglades National Park into a highly-controlled agricultural water supply and delivery system. As a result, the rich littoral zone of diverse island and shoreline vegetation has been diminished. The artificial enrichment of the lake by agricultural wastes has increased ten to twenty fold, thereby aging the lake through a process of chemical overloading called eutrophication. And as the Water Resources Atlas of Florida candidly puts it, "the ability of the system to store and release excess waters to Lake Okeechobee and its ability to purify the runoff it receives has largely been destroyed. . . . A hydrological yo-yo situation [now exists] in southern Florida . . . [and its] effects on their natural plants and animals . . . are ghastly and tragic."

Jono and I had canoed up to a place where the channel was clogged with water hyacinths and water lettuce. What we would see once we portaged this impasse gave us a glimpse of the lake's tragedy, among other things.

We came around a bend, and there, on the lakeward side of us, we spotted a drowned remnant stand of custard apples, *Annona glabra*. The trees stood nearly leafless, in about a foot and a half of water. They are capable of enduring water over their roots for a long "hydroperiod" each year, but may tolerate or even require unflooded soils for brief periods. Formerly, this particular stand of one hundred or so trees had been much more extensive, but it had been permanently flooded when the Water Management District raised the lake levels several years ago.

Jono and I decided to hop out of the canoe, wade over, and investigate the trees to see if they held any custard apples, a popular edible fruit in the tropics. About fifty yards past a muddy clearing where three baby gators were resting, we found a break in the floating smartweed. There, we safely moored the canoe and waded toward the trees.

As Jono sloshed up to the first custard apple tree, I noticed something floating beneath it. I did a double take. There was a single floating gourd.

We picked it up. It was a slightly punctured, waterlogged, hard-shelled fruit with fermented pulp and seeds oozing out the side. What a battered, soggy piece of evidence that a species still existed!

We scoured the vicinity, looking for leaves, tendrils, vines, or their remnants. We dodged floating fire ant colonies, and retreated from spots where snakes swam out in front of us. Each custard apple and willow was checked for evidence that it had served as a trellis for climbing plants other than the ubiquitous moon vine. But no such clues could be found. Perhaps the gourd had floated in from somewhere nearby.

We returned to our canoe, paddled past the wallow where the baby gators had been sunning themselves, and then we floated around a bend. Jono spotted the last few trees in the custard apple stand, and we decided to spot-check them. We waded through more hyacinths, water lettuce, and smartweed toward the outlying trees.

"Is that one dangling under the tree?" I asked Jono. Just then, he spotted another gourd floating under the custard apple. We investigated further. Both of these gourds were still attached to their skeletonized vines, and were not waterlogged like the one we had seen earlier. One, in fact, was dry as a bone, dangling a foot or so above the water's surface. Although we could not trace their vines back to the main stem or root of a withered annual plant, it was clear that they had grown up under this isolated custard apple. In a "land" where water levels could rise at any time, this tree-climbing vine had advantages over those that only sprawled over the ground.

Still, the sight of two gourds ornamenting a dwarf custard apple was a meager discovery compared to the vision of custard apple forests that Florida Cracker Lawrence Will had seen more than a half century before: "When these woods were in their prime, exploring their shadowy domain was an experience you'd not forget. . . . [Y]ou might walk for miles and scarcely glimpse the sky. . . . Gourd vines with their green pendant fruit, looping and lacing from branch to branch, were less a barrier than those yellow strands, tough as piano wire, spun by enormous brown and yellow spiders."

Ethnobotanist John Harshberger added that the Lake's custard apple forests formed such dense shade that at any hour "a twilight pervades the solitudes."

From the descriptions of Will and Harshberger, I could imagine the custard apple forests as they had been when they covered 32,000 acres on the lake shoreline and the islands. They consisted of thirty- to forty-foot-tall trees with overlapping canopies, woven shut by a tangle of gourd vines and lianas. This created an understory microenvironment distinct from anything else near the lake. Jono and I could see no such forest now, only a scattering of ten- to fourteen-foot trees, and a couple of gourds hanging on for dear life.

Realizing that the demise of this useful fruit had happened within his own lifetime, pioneer Lawrence Will lamented the changes his contemporaries had made on the southern rim of the lake: "And so now today, of all those millions [of custard apple trees] that once ringed the lake, how many do you suppose there are left? On the marshy western shore of Torrey Island, lonesome, scraggly, and discouraged, you may find a dozen stunted trees, sorry examples of the former forest, and Sherlock Holmes himself couldn't find a dozen more."

Neither Jono nor I compares well to the world's most famous sleuth, but we did encounter upwards of two hundred custard apple trees in various remnant habitats. Regardless of their height, most appeared to be approaching senility. They were usually bare of leaves. Many branches appeared dead. Water stood to such a height on their trunks that a dry gourd could not seed itself in unsaturated soil beneath their shadows, at least not under present water management practices. The chances for their reproduction were low. If the custard apples themselves were not reproducing to any extent, one could bet that the Okeechobee gourds would have trouble reproducing. Their fates seemed intertwined.

Upon returning to the canoe, Jono was somber. He was pondering this dilemma, as if trying to paddle through a choked, disrupted landscape: "There ought to be just one place," he opined, "where people could still go to see custard apples and gourds and moon vines together, the way the whole south rim of Okeechobee used to be. I think that's the least they could do, to try to keep one custard apple swamp alive somewhere."

IV.

"The least they could do." By "they," Jono meant the South Florida Water Management District, because he spoke with them on the phone soon after we returned to our "wilderness campsite." The District owned the islands where the gourds were found, having purchased the land after establishing high water level policies that threatened to flood the farmers who had cleared the island interiors. Despite the fact that the District engages more natural scientists in work around the lake than any other agency, its personnel had never systematically sought out the gourd, nor done any review of its rarity in relation to their management practices. We decided to make the gourd's presence known. We hightailed it to West Palm Beach the next day.

Once we had found our way through the bureaucratic labyrinth, we were introduced to a couple of the District's environmental scientists in a back office.

"We were looking for the Okeechobee gourd and custard apples on the west side of the islands," Jono began, "and wonder if you have aerial photos so that we can pinpoint where we went."

"Oh, you were looking for them from an airboat," one of the scientists commented.

"No, we canoed," Jono replied.

"Oh, then you never reached the custard apple stands up there, that's too bad. The channels are really choked with hyacinths, aren't they?"

"No, we got there all right," Jono said.

"We just had to drag the canoe for eighty or ninety yards through the hyacinths," I added.

"These boys are not candy-asses!" the senior scientist interjected, bemused.

"Well, it's great that you reached the custard apples," the other scientist said apologetically. "But I suppose you didn't see any gourds . . ."

After confirming that we had found the gourds, we changed the subject to water levels. The District's policy is to maintain the lake's level between fifteen feet and seventeen and a half feet above sea level, in order to provide storage of water supplies for agricultural irrigation and municipal needs. During infrequent drought periods, the lake level drops several feet below the fifteen foot minimum, as it did the last time the gourd vines had been seen in 1981. During our visit, however, the lake level was just above the minimum, at 15.2 to 15.3. Given the amount of water already in the gourd's habitat, this meant that the lake level would have to drop no less than fifteen inches below the District's minimum if the gourd seeds were to find unflooded ground for germination. Drought years were thus more helpful to the gourd's survival than the District's own water policies.

I knew that the state of Florida listed the Okeechobee gourd as endangered, but I assumed that there were other rare species that were just as much affected by artificial lake level fluctuations. I was curious to know if the District had to take into account the effects of water management on any other animals or plants rare enough to be protected by federal or state agencies.

"In terms of your water management policy, are there any threatened species that concern you more than the gourd?" My question apparently touched a nerve:

"*Any* other resource is more important to us than that gourd!" one of the scientists quipped. "We try to take into account the Wood Storks and the Everglades kites. But no, no one has ever thrown an endangered plant at us!"

We left some of Jono's posters behind, to stimulate further interest in the gourd. One of the men incidentally offered us directions to other custard apple swamps of which he had heard. Though the meeting was cordial, it was clear that the District's environmental scientists had worries other than the fates of hard-shelled gourds and soft-fleshed custard apples.

V.

On the way back to the lake to search more custard apple swamps, I was struck by the omnipresence of smoke in the sky. The Everglades region oftentimes has a pale, powder blue sky, carrying vague wisps and puffs of clouds during the dry season. But as common as the clouds are the dirty gray smoke plumes rising over sugar cane fields.

Along the highway, we stopped where one old cane field had just been lit. The fire crashed and cut through the stubble with a cracking sound not unlike that of the machetes we had heard in another field nearby. The flames crested ten to twenty feet above the ground, and were enveloped in billows of smoke. After the first rush of fire across the field, the cane debris and even the sawgrass peat soil beneath it would smolder for hours.

This burning reminded me of what had happened to most of the custard apples that once grew around the lake. While the remaining custard apple swamps face too much flooding today, the historic threats were draining and burning. The reason, as Lawrence Will told it, was simple: "The custard apples made one bad mistake. They picked for a place to grow the finest soil in all the world, and when farmers found this out, then it was good-bye to the custard apples."

The soil of which Will spoke is technically called Custard Apple Muck. It once formed a natural levee on the southern shores of the lake, allowing a diffuse spillover into the Everglades during the rainy season. This earthen bank is a complex layering of rich, dense, black granular mucks derived directly from the custard apple swamps, and fibrous, brown-gray peats from poorly decomposed Everglades sawgrass. When farmers originally burned the swampy forest after draining it, some say that the exposed peats caught fire and smoldered for weeks.

John Small, the botanist who named the Okeechobee gourd, observed that "it will not withstand fire." He surmised that "it is less resistant to fire than the cultivated pumpkins, for the flesh of the fruit is much thinner, and when dry, together with the woody shell, readily burns and destroys the seeds." This, Small added, posed a significant problem, since "at least ninety-five percent of its former geographic area has been reduced by fire to ashes and smoke."

The torching of vegetation to clear it is only one kind of burning. A more insidious kind of burning has occurred on the southern rim of the lake, without a match being struck or tinder being kindled. This is the slow oxidation, followed by drying and shrinking, of organic soils.

The biochemical burning or oxidation of peat and mulch south of Okeechobee has affected about 1760 square miles of Everglades soils. About twenty-five percent of these were so shallow that it did not take too many years of plowing and burning to completely dissipate them. The remaining 1100 square miles in agricultural use may not survive another two decades given the current rates of soil depletion. Lake Okeechobee, then, has not only lost most of its gourds and its custard apple swamps, but also the organic soils that once supported these plants. During the last seventy-five years, agricultural experiment station staff have kept a benchmark by which to measure this loss of soils in the Everglades region. South of the lake a few miles, ground levels have dropped seven feet below this benchmark since 1912. Soils that were once at least eleven feet deep are less than four feet deep today.

What forces have worn away the soil, and exacerbated the extirpation of the gourd, a symbol of the earth for many peoples around the world? While the last century of exploitation has fully exposed the land to this suffering, I sense that a series of cultural disruptions over several previous centuries set the stage. If there had been any significant cultural stability in the Okeechobee region, it is unlikely that devastation of this scope would have been tolerated.

How long this instability has been pervasive is not known. We do know, however, that between 500 B.C. and 1545 A.D., prehistoric cultures in southern Florida gradually developed considerable knowledge about native plants and animal resources, managed some of the wetlands near the lake, and carried on far-reaching cross-cultural trade. One of these cultures, the Calusa, built permanent homes, temples, and burial grounds on raised earthen mounds near Lake Okeechobee, and adjusted their food-gathering habits to fluxes in the lake's levels. Fish, shellfish, turtles, alligators, deer, birds, cabbage palms, tubers of the coonti, fronds, and wild grains were among their local foods, and others were no doubt traded in from neighbors. One neighboring group was called *Tocobaga* for their home ground, meaning "place where gourds grow."

The Calusa burned to manage patches of cabbage palms and other edible plants, but used other methods for producing food as well. By excavating shallow marshes to build up long earthen beds, they created new environments on which to intensively grow crops. These farm plots, sixty to six hundred yards long, were not unlike the raised fields of the lowland Maya, who shared numerous other cultural traits with the Calusa. The Mayan region of southeast Mexico is also home of the *Martínez* gourd, the other subspecies of the Okeechobee gourd. Geneticists Robinson and Puchalski have suggested that the Okeechobee gourd must have been carried from Mexican cultures during prehistoric times, perhaps by boats that travelled considerable distances along the Gulf coast, or that hopped from island to island across the Caribbean. They note that the gourd's "hard rind and good keeping ability would make it possible for the fruit to be transported from Mexico to Florida, even with the very long time such a trip must have taken. . . . The extremely bitter flesh of the species precludes it from being used as food, but the seeds are edible and nutritious and the fruit could also have been used for its detergent quality."

Calusa settlements may have received the gourd through their extensive trade networks, but these networks may have also hastened their downfall. Once European diseases arrived and infected one Florida Indian settlement, they rapidly spread to others. It is likely that these diseases, more than battles, slavery, or food shortages, reduced the Calusa down to a few thousand survivors. Then, the surviving Calusa dispersed, some settling in the Keys and perhaps even in Cuba. Others at first fought with, and then became aligned with, the southward-moving Creeks in the early 1700s. Some historians suggest that as Creeks from Alabama and Georgia readapted to the Everglades environment, the remnants of Calusa knowledge helped them make the transition. Along with the Hitchitee-speaking Mikasukees, these people became known as the Seminole, from the Creek term, Iste Semole, meaning "free being" or "runaways."

As Creeks, Mikasukees, and Blacks continued to run away from Anglo-American soldiers and settlers intent on enslaving them and seizing their land, the Seminole amalgamation retreated further in the Lake Okeechobee glades and Big Cypress Swamp. There, they raised small patches of corn, bananas, root crops, squashes, and gourds. Around Indian Town and other Seminole camps, early observers reported the large butternut-like Seminole pumpkin and a diminutive gourd. Both of these had vines that would climb into surrounding trees, where their fruits would sometimes hang beyond reach of the gardeners who had planted them. These hard-to-reach ones, a Seminole quipped, "were for the 'coons."

The Okeechobee gourd and the Seminole pumpkin species, *Cucurbita moschata*, can produce partially fertile hybrids if cross-pollinated by bees or by plant breeders. Florida's economic botanist, Julia Morton, has commented to me that she "can't help wondering if the Seminole pumpkin may not be a hybrid between *Cucurbita moschata* and the wild Okeechobee gourd, or at least . . . accidentally crossed with it. [G]rowing the vine on trees seems a practice peculiar to the Florida Indians. . . . [I]ts sturdiness and self-perpetuation, together with the great variation in hardness of rind, size, form and color [make it unique]."

Morton has observed that Seminole pumpkins cultivated in Florida infrequently carry wild traits: extremely hard rinds, green and white striping, and a strange unpleasant aftertaste, perhaps from low doses of bitter cucurbitacins. These are all characteristics that the gourd could have passed on to *moschata* pumpkins after the latter were introduced from Mexico or from the Caribbean in early historic times. Still locally popular in Florida, the fine-tasting Seminole pumpkin may owe some of its distinctive qualities to the gourds.

It was fortunate that both the gourd and the pumpkin had the capacity to persist in a feral state around Indian camps for a time after the camps were abandoned. Warfare frequently forced the Seminole to temporarily abandon favorite camps. Between 1818 and 1856, the United States government attempted either to remove or to kill all Indians remaining in Florida.

Like the guerrilla fighters of modern times, the Seminoles of the nineteenth century would resist, attack, and then retreat into the roadless swamps of southern Florida. From the first territorial governor of the state, General Andrew Jackson, to business-oriented Governor Brown, who thought he could buy them out of the Everglades, the Seminole were considered obstructions to progress. In 1852, Brown bluntly stated that "the most interesting and valuable part of our state . . . is cut off from any benefit to the citizens and sealed to the knowledge of the world, to be used as a hunting ground for a few roving savages."

Despite the implication that the Seminoles were merely savage hunters, the whites on occasion incited conflict by demolishing Seminole plantings of food crops. In 1855, a surveying expedition "raided Billy Bowlegs' garden . . . and tore down his highly prized banana plants. . . . On Bowlegs' return to his garden he flew into rage when he saw the damage done . . ." The next morning, he killed two whites.

The end of the battle was near, however. Within a year, all but one hundred Seminoles agreed to move out of Florida for money and for the promise of land. Tiger Tail, an old Seminole resister, committed suicide on the way to New Orleans, and within a year and a half, Billy Bowlegs himself was dead. Some of the transplanted Seminoles later rejoined their kin who remained hidden in the Everglades, but their battles with the whites were over by the time the Civil War began.

At the conclusion of the war, the Seminoles and southern whites around Lake Okeechobee were living as neighbors without major conflicts. But northern carpetbaggers quickly saw that the undeveloped lands around the lake might make them millions. In 1881, a northern industrialist, Hamilton Disston, was granted the rights to drain overflowed land adjacent to Lake Okeechobee in return for half the area reclaimed. He soon became the largest landowner in the United States, and began the environmental disruption of the lake's watershed that has spelled doom for wild gourds and sustainable agriculture alike. By 1884, his companies had drained over two million acres near the lake.

Although Disston died before his project was completed, his schemes brought in land speculators and cash-croppers from all over the eastern states. By 1910, the government was involved in opening new agricultural lands by building new canals, roads, railroads, and towns.

Destined to be a fertile vegetable growing area, the southern rim of Lake Okeechobee was drained, plowed, and planted. At first not all the vegetable crops survived; the mucks were rich in organic matter, but deficient in certain essential minerals. "Reclamation disease," also known as "muck sickness," continued to ruin certain crops until agricultural scientists determined which fertilizers should be applied to offset nutrient deficiencies in the soil.

Once artificial fertilization was accepted, then Everglades agribusinessmen were quick to accept other modern technologies: drainage techniques, pesticides applications, and vegetable processing for canning. Lake Okeechobee's fringe became one of the first vegetable growing areas in the country where airplanes were routinely used for spraying and dusting. Aerial seeding and even low-flight air stirring to protect crops from frosts were also attempted. With few immediate social or environmental constraints working to slow or offset modernization, one Everglades county adjacent to the lake soon led all counties in the U.S. in the cash value of its crops.

The variety of crops grown, for a while at least, was great. Snap beans, celery, potatoes, lettuce, peanuts, dates, citrus, and cane consistently provided the highest cash returns, and until the sixties, a fifth of all production was spread through a diversity of other minor vegetables. A few diversified family farms became known as "thousand acre salad bowls." Squashes and gourds, however, were never among the significant crops.

By World War II, corporate farming was developing on such a scale that "native labor"—local blacks and Seminoles—could not meet the demand. Migrant workers were brought in from as far away as the West Indies and Mexico. The conditions of the barracks and bathrooms that companies provided were notorious nationally, yet workers had to pay a significant portion of their wages to use these facilities. Everglades agriculture had begun to degrade people, much as it degraded the Florida soil and watershed.

When Castro took over Cuba, sugar cane, an Everglades crop since Disston's time, became a strategic resource that the muck and peat of Florida's custard apple swamps could provide to the entire United States. However, farmers soon learned that as they burnt the cane stubble, the peat soils would burn as well. The old paved roads through canefields now dip and buckle due to the resulting subsidence of the soils adjacent to the roadbeds. The switch to cane monoculture has simply hastened the drop in soil levels.

Profitable cane production does not necessarily go hand in hand with social welfare, either. Today, the run-down company-owned barracks and buses for Cubans, Haitians, and Puerto Ricans are not much improved over those that sparked federal investigations in decades past. Though sugar may be sweet and white when packaged, it leaves puddles stained and stinking with cane char and diesel oil in its wake.

Where the smell of cane smoke is in the air, there is often the feel of cultural disruption on the ground. Cheap alcohol and prostitution may not be more common on Okeechobee's southern shores than elsewhere in the United States, but they are still a blatant reminder of the ripped fabric of life there. In recent years, however, AIDS has become the local indicator of social distress. Belle Glade, on the lake's southeast shore, is reputed to have the highest density of AIDS victims in the U.S., outstripping San Francisco and New York. It is but one of the many common afflictions that white Crackers, Southern blacks, Haitians, Cubans, Seminoles, and Puerto Ricans now fear.

VI.

Jono and I finally left the lake's southern shoreline to search roadsides and canal banks leading away from the lake. Within the last decade, gourd "escapes" had been sighted on ditchbanks some distance from the lake. These out-of-place plants must not have persisted. We spent hours looking for other stray gourds, to no avail.

I guessed that weed control along ditches and fields away from the lake would be severe enough to terminate the growth of any new gourd vines. This guess was reinforced by a talk with the grounds manager of a local recreation park adjacent to canals and cane fields.

"Sure, there used to be vines like them here in your picture," she said, examining Jono's poster. "They were over on that edge at one time," she said, pointing.

"I suppose there's a lot of different kinds of vines here," I replied. "This kind has hard-shelled fruit that get to be the size of baseballs. Seen anything like that?"

"No, we don't let anything get that big before we cut it out. If we can't cut 'em, we use diesel oil or Doomsday. We try to get them out as fast as we can spray 'em."

With that, Jono and I left to drive westward, back past the signs that joked, "It never hurts to raise a little cane." As we veered away from the lake, Jono pointed out where more native vegetation was being cleared away. Small patches of prairie wetlands, overlooked during earlier boom periods, were being drained for pasture or for cropland.

"Well, I suppose they'll tell us they can't live on scenery alone," Jono sighed sadly.

Jono Miller had put his finger on an attitude that affects thousands of rare plants and animals. Most modern farmers assume an inalienable right to convert wildland habitats to "productive space." If they are cognizant of the wildlife they displace, they presume it will survive elsewhere. They are in the business of production, not scenery appreciation.

A cane farmer, from this strictly economic orientation, should not be constrained from clearing a custard apple forest full of hanging gourds. After all, even if custard apples and gourds were useful to someone else, he gains no financial benefit from them. He has been encouraged to think of his value to society not in terms of land stewardship, but in terms of the crop production for which he is paid.

Following the logic of self-interest, a farmer can rationalize away any sense of community responsibility. What if his land maintains a gene pool of a crop relative, potentially valuable to those who grow that crop in some other corner of the world? Will those farmers pay him to keep part of his lands wild enough for the gourd or some other rare plant, to offset the loss of income otherwise gained from growing crops? Even if he lets a plant breeder come and collect some gourd seeds, then the breeder and farmers elsewhere, not the farmer near Okeechobee, benefit economically from subsequent advances in crop improvement.

The Okeechobee gourd may in fact have value to other farmers worldwide. It is a source of genetic resistance to powdery mildew, bean yellow mosaic, cucumber mosaic, and tobacco ringspot virus. Its seed oils are admirably high in linoleic acid, a polyunsaturated fatty acid. Of all the *Cucurbita* tested so far, it has the highest content of bitter cucurbitacins, which naturally attract corn rootworms and cucumber beetles. This makes it an excellent candidate as a trap crop or a supplier of beetle larvae attractants for integrated pest management schemes. Such schemes may save farmers millions of dollars in yield losses and pesticide costs for crop plants more vulnerable to these larvae than are the gourds.

The qualities of Florida's remaining Okeechobee gourds are all of *potential* value. Their closest relatives, the *Martínez* gourds of Mexico, have already proved the point. Genes for powdery mildew and cucumber mosaic virus resistance—comparable if not identical to the genes of the Okeechobee gourd—have already been transferred from them to commercial squash cultivars. This powdery mildew resistance is now protecting butternut squash crops around the world. In addition, it is being bred into the most widely-grown squash and pumpkin species, *Cucurbita pepo.*

Still, should this convince cane farmers to manage the lake's irrigation system differently, so that gourd habitats will not be inundated every month of the year? Should farmers feel obliged to protect wildlands on their own properties, wildlands that are potential habitat for gourds? If squashes are no big piece of the Okeechobee pie economically, why should cane farmers concern themselves with a few gourds?

In a stable cultural community of local farmers, perhaps such questions are never asked; wildlands are maintained, because worship of the dollar does not drive every farmer toward plowing up all of his available land. But for Florida's sugar cane farmers, involved in an international economy, a global responsibility is warranted, for they themselves are dependent upon genetic resources from other lands on this planet. Ask sugar cane farmers and mill owners around Lake Okeechobee where their industry would be today if gene transfer had not rescued it from the mosaic virus that spread through the region after 1914. Before plant breeders could find sources of resistance to sugar cane mosaic, the virus had devastated three-quarters of the crop. Over $100 million was lost to this disease in the southern states during the epidemic.

Fortunately, genes for resistance were found in wild sugar canes from Indonesia and other countries of tropical Asia. Three disease-resistant varieties were bred at the Canal Point Breeding Station not far from the lake. They helped revive the sugar cane industry in south Florida, which went on to gain cane yield records for the United States.

Today, modern agriculture is expanding in Malaysia, Papua New Guinea, and Indonesia, where it is wiping out smaller-scale traditional farming as well as the valuable gene pools of wild cane still found in the rural landscapes there. Wild *Saccharum* species such as those now being lost have been used for developing resistance to bacteria, rats, and fungal root rot in recent decades. As geneticist J. D. Miller told Robert and Christine Prescott-Allen, "if no germplasm from wild relatives had been used, there would probably not be a viable sugar cane industry any place in the world."

Since Florida sugar cane was revived through the use of Indonesian sugar cane gene pools, it is equitable for Everglades cane farmers to consider the value of the gourds in their watershed for the rest of the world. The global economy is now dependent upon such humble plants, which will survive only if their value is recognized in their homelands.

VII.

My preoccupation with lost gourds and spent soils subsided as we passed one last time through the Brighton Indian Reservation of the Seminoles. Traditional *chickees* or ramadas topped with cabbage palm fronds were a refreshing sight after too many smoke-stained sugar mills. Seminole grocery stores and craft shops were loaded with Indian drums and tomahawks for the tourists, but the few Seminoles I encountered still seemed to carry a distinctive identity with them. As Jono and I pulled into tribal headquarters to leave a gourd poster, we saw a woman leave there wearing "traditional" dress—a blouse and full skirt made from strips of brightly colored cloth, stitched together into a dazzling collage.

We walked into the receptionist's office at tribal headquarters.

"Is it okay to leave this poster here?" Jono asked.

"Sure, go 'head."

"We were wondering if there's anyone around here who grows squashes or gourds," I asked.

"No, I can't think of anyone who would be able to tell you anything about something like that," the Seminole woman replied. Case closed.

"Thanks."

We walked back to the pickup. Jono was grinning from ear to ear.

"Gary, did you see what that woman had on the table next to her desk?"

"Nope. What?"

"A gourd . . . or a small pumpkin . . . It was striped with a slight neck."

We couldn't be sure what kind of cucurbit it was. Perhaps it doesn't matter. But as I left Seminole country, I felt there was still a chance that someone near Lake Okeechobee had a soft spot in her heart for gourds.

(Postscript: Since the rediscovery of the gourd and follow-up research, U.S. Fish and Wildlife Service botanists in Florida have vacillated over federal listing of the Okeechobee gourd as an endangered species. This action has been vigorously supported by many gourd enthusiasts and genetic conservation specialists. However, its ultimate effects on the management of water levels, and the probability of gourd survival in custard apple swamps, have not yet been determined.)

CHAPTER TEN

Drowning in a Shallow Gene Pool:

The Factory Turkey

I.

Turkey under glass. I peered through the pane at the mass of mummified tendon, bone, and feather. These desiccated remains represented the only domesticated bird species bred in prehistoric North America. Forty years before my own descent into Canyon de Chelly, a park ranger had uncovered this male turkey intact within a Basketmaker cave. He carried it out, and the National Park Service has guarded it ever since.

Estimated to be 1700 years old, this specimen of the Small Indian domestic breed eventually made its way into a museum case displaying corn, lima beans, and clay bowls. For most museum visitors, its frayed feathers and withered flesh are an unappetizing mess. They quickly pass by the desiccated turkey, eager to look at items more aesthetically pleasant: petroglyph replicates; decorated pottery; or handspun weavings.

What first drew my attention as I gazed at the bird was that its feet were folded, as if in prayer, the claws of one wrapped into the other. My vision moved up from the scaly toes to scan the white chevrons barring the gray-black feathers, a good three dozen primaries and secondaries still hanging onto the wings. Then I saw the hole gouged in the upper breast.

Looking hard through the reflections on the glass, and the shadows behind it, I could see something within the hole which makes this bird both a natural and a cultural artifact: a corncob.

Only then did it dawn on me that the tom turkey mummy lacked an essential feature: its head. A strand of vegetal cordage had been tied around its neck, and the head had been ceremoniously severed. After the turkey was beheaded, apparently a corncob was stuffed into its gullet, and the entire bird was stuffed into a hollow in the sandstone cliffs near the mouth of Canyon del Muerto.

II.

It struck me as funny that most prehistoric turkey pictographs have no heads. A few hours after peeking at the beheaded bird in the little museum, I examined the Anasazi rock art on the sandstone flanks of Canyon de Chelly and adjacent Canyon del Muerto. The massive pink, beige, and peach cliffs are stained naturally with cobalt blues and rich red-browns, and also bear hundreds of petroglyphs and pictographs made by the prehistoric Anasazi and historic Navajo and Hopi. Glyphs of gobblers by themselves, or sitting on the heads of humans, are the most common representations of bird life in canyon country rock art.

At Ledge Ruins, where many turkey bones and feathers have been excavated over the years, a Navajo guide showed me the blushing sandstone wall where the cream-colored crescents were drawn. Early amateur Anglo interpreters of Southwestern Indian graphics identified these images not as birds but as baskets or "lazy crescents." Perhaps these rock art students were lazy themselves, for if they had only asked contemporary Indians, they would have correctly identified the icons as turkeys much earlier.

Anasazi artists, as scholar Campbell Grant later surmised, painted their gobblers as bicolored birds, with the heads and necks often done in a reddish tint that quickly faded. Only at a few naturally-protected sites does this usually fugitive head color persist. In these preserved paintings, the reddish tints of the turkey's knob-like caruncle, snood, and wattles add an uncanny accuracy to the images.

At Antelope House not far from Ledge Ruins, gobblers are infrequent among the Anasazi petroglyphs left on the towering six-hundred-foot sandstone walls. Nearby, in the pueblo, there remains an eight-inch-deep layer of turkey dung, eggshells, and bones of poults. This layer was deposited over the half century or so following 1125 A.D., during a period of breeding for feather color diversity there. The Antelope House community restricted distinct turkey strains to different areas of one particular plaza behind the pueblo's rooms, presumably to enable controlled breeding for plumage variants. The Anasazi had plenty of time then to observe turkey behavior, and accordingly Grant feels that their turkey drawings are more realistic than any of their other animal drawings.

Turkey beaks, where they persist in Anasazi paintings, have the correct downward curve to them, and the fleshy snood above the beak is sometimes shown in its excited state, enlarged as it is during courtship. "Wings are never indicated," wrote Grant, "but an effect of flight is achieved by extending the neck of the bird forward and angling the feet back." Having once seen a flock of Merriam's wild turkeys suddenly leave the ground and alight in some ponderosa pines, I can attest that the Anasazi somehow captured the essential motion—or commotion—of turkey flight.

Despite its powerful spurts of flight over short distances, the turkey is the one bird considered by Puebloan cultures to be more a creature of the earth than of the sky. More properly, perhaps, it is a symbolic link between this life on earth and the spirit life. Archaeologist-artist Polly Schaafsma has verified that some Pueblo Indians today view turkeys "as an intermediary between mountain water sources and the rain clouds" and as companions of the dead "who must return to earth before rising as clouds to the spiritual realm."

This symbolism did not develop recently, for centuries-old mummified turkeys and bundles of their feathers have been found buried within human burials in the Southwest. At Casas Grandes, the fallen prehistoric trade center that linked Mesoamerica with the arid hinterlands of the Southwest, most turkey remains were found in the House of the Dead. Schaafsma believes that the prehistoric practice of wrapping corpses in turkey feather robes was "a means of assisting the dead on their spiritual journey," for turkeys are still regarded as teachers and helpers for such passages by Puebloans today.

III.

Those "teaching turkeys" were a far cry from the dumb, broad-breasted domestic breeds that most Americans have known since World War II. The authors of a National Research Council publication on small livestock had difficulty retaining their usual style of objectivity with regard to modern turkeys: "The commercial birds are not considered intelligent. They often have to be taught to eat and drink, and they become lost easily. [They have] lost many abilities of survival in the wild. . . . Modern turkey breeding has been so dominated by selection for increased size and muscling that commercial turkeys can hardly walk. . . . These highly bred 'freaks' are adapted only for complex intensive production, and have to be raised with the strictest care."

Turkey breeds have become increasingly overspecialized since World War II, when the government encouraged small farms growing mixed crops to shift to exclusive production of a few cash crops and breeds suited to the packaging plants built in their localities. In his historic novel *Leaving the Land*, Douglas Unger traces the fall of a diversified Dakota family farm as it converts to turkey monoculture. As her father puts all his economic eggs in one basket, Marge Hogan finds that the pale and pluckable White Holland is a bird altogether different from the hardier heirloom breeds of turkeys they had once kept in the barnyard:

> She vaccinated thousands upon thousands of feeble balls of gray down she could hold in one hand. She put young chickens in the pen with them. Turkey poults could only learn to peck by watching chickens. Without chickens, they stood in piles of grain and died of starvation. . . . Then whenever it threatened to rain, Marge and her father

had to rush out into the fields to herd their turkeys back into the barn. . . . [A]fter the slow realization that it was water, not food, moving on the ground . . . that entire flock in an incredibly orchestrated movement raised their scraggly necks from the ground, tilted their bald heads to face the sky, and opened their beaks wide to the falling rain until they drowned.

The first turkey factory that I ever visited was near Goshen, Indiana. Many farms with mixed crops and livestock persist in the Amish and Church of the Brethren communities there. However, a few families have switched from keeping free-ranging turkeys in fields with portable shelters to raising turkeys intensively in stationary pole buildings. Constructed from telephone poles and steel siding, equipped with heaters, fans, and specially-prepared feeds laced with antibiotics, these artificial environments should theoretically keep turkeys under optimal conditions for rapid growth. Thousands of turkeys are raised together in a single pole building, and they behave as if guided by a single, simple mind.

I recall my horror as I entered a pole building early one warm spring Sunday morning and was met with the stench of sick, dead, and dying birds. They had probably been struck by a protozoan infestation, which, when fueled by overcrowding, can quickly decimate whole flocks. The droppings had become bloody or discolored, and the young poults were overcome by feebleness. My friends pointed to where dozens of sick birds had fallen into the feed troughs and suffocated. We sifted through feed and pulled the dead birds out by their feet.

As the waves of shrieking turkeys crowded around us on the floor, weakened ones were being trampled beneath them. We hauled the newly-dead over to a fifteen-foot-deep hole full of turkey corpses, and tossed them in. Leaving the pole building, fresh cool air never tasted so sweet.

Protozoans such as those causing blackhead and coccidiosis may infect the guts and bloodstream of wild or domestic fowl, and hybrids between them. These epidemics become far more prevalent when the birds are cooped up. Where domestic turkey densities rise to thousands per acre, chances are high that the microbes infecting the droppings of one sick bird will reach the others. Wild or part-wild hybrid turkeys are more likely to contract the disease if they are penned in on game farms. The same is true if large flocks range into fields where chicken manure from contaminated fowl has been spread as fertilizer.

But it is not normal for a large flock of wild turkeys to be confined into a small area. It may be true that where food is abundant on farms or in forests you sometimes find flocks of a hundred wild turkeys foraging together on a few acres for several days. However, those that freely range on wildlands seldom exceed densities of ten to twenty birds in three acres. A single flock of wild turkeys may travel several miles a day while foraging. Its feeding is not at all concentrated, as is that of commercialized domestic droves.

While some of the vulnerability of modern domestic birds comes from confinement in artificial environments, turkeys also suffer from genetic vulnerability. Most of the 124 million turkeys in the world today belong to a handful of uniform breeds, all of which are decended from a few Mexican domestics taken to Europe in the fifteenth century. The Atlantic Ocean served as a bottleneck for the turkey, for its gene pool became shallow on the European shores. From the small founding populations they had to work with, Europeans selected for turkeys with deeper bodies and heavier flesh, particularly on the breasts and thighs. Modern birds have shorter, thicker legs and necks, smaller brains, and larger secondary sexual characteristics such as wattles. White-feathered breeds, which appear more cleanly picked at early ages when pinfeathers are still a problem, are favored by factories.

This kind of genetic streamlining has happened with numerous domesticated plants and animals, but in certain modern breeds of turkeys, it has resulted in a near-fatal flaw. By packing so much of a tom turkey's meat mass onto his breast, he can hardly make cloacal contact with a hen during mating. As Lawrence Alderson has written, " . . . The most bizarre example of the adverse effects in domestic lifestock is provided by the Broad-breasted White Turkey. This breed has been so overdeveloped for meat production that it is unable to mate naturally because of its shape. It relies entirely on artificial insemination to reproduce itself."

IV.

When did turkeys begin this inexorable strut toward an evolutionary dead end? What became the point of no return, at which selected variants deviated so far from the wild type that survival in natural habitats was no longer possible?

Pre-Columbian turkey breeders may have lacked the technology to inseminate hens artificially, but they were fully capable of selecting aberrant forms that differed from wild types. The basic science and the skills of animal breeding were already in hand. According to zoologist John Aldrich, one turkey specimen from a prehistoric pueblo "has its neck fully feathered up to the skull, unlike any known wild variety of turkey." Feather-necked birds were common among the Small Indian domestic breed, but are extremely rare among wild and domestic turkeys today. Turkey mutants lacking certain pigments in the feathers and feet crop up infrequently among the Southwestern Indian breeds, but more so than in the wild.

Considering these strange birds, we cannot deny that prehistoric peoples had the ability to keep "unnatural" forms alive once found, or even to select for that which might be maladaptive or detrimental in the wild. Prehistorically, Native American minds had the same capacity to deviate from nature as we do.

Why, then, did Indians favor forms other than those approaching the Broad-Breasted White Turkey? How is it that some of the Indian turkeys retained the capacity to become feral, but modern breeds cannot do the same today? If we are ever to become able to answer such questions, we will have to consult the bare bones in prehistoric ruins, and consider the fleeting descriptions of turkeys and their keepers left us by early European visitors to the Americas.

V.

First there were the wild turkeys, bright in mind, quick in movement. "Probably no other bird or mammal excels the turkey in alertness," wrote New Mexico ornithologist J. S. Ligon. "It can instantly detect the slightest movement of an object in the scope of vision. . . . Even the shriek of a chipmunk, chatter of a chickadee, or the scolding of chickadees or tiny bushtits is sufficient to cause all turkeys within hearing to snap to attention. . . . If the danger proves real, the skill and speed with which all disappear is astonishing . . ."

Before humans began breeding turkeys, natural selection had already diversified the gallinaceous genus *Meleagris* into three wild species strewn across the Americas. One of them, *M. crassipes*, was widespread during the Pleistocene and persisted at one semi-arid site in North America as recently as 3000 to 6000 years ago. Cultural remains are seldom if ever found with this Pleistocene holdover, although it must have been contemporaneous with early Holocene hunter-gatherers. Medium-sized but quite distinct in shape, fossil bones of this odd bird show almost no sexual dimorphism. That fact baffles zoologists, because the differences between the sexes are so pronounced in the other two turkey species.

In tropical Mexico and Belize, the Oscellated Turkey is a brilliant bird with orange-red knobs and warts on a bare blue head, and a blue-green "eye" appearing to look back at you from its barred gray tail feathers. *Meleagris os-cellata*'s range was restricted to the Yucatan peninsula, and it was intensively trapped and eaten by the Mayans there in prehistoric and historic times. Yucatan, in fact, was called "Land of the Deer and (Oscellated) Turkey" by the Mayans. Thousands of bones of these two game species are found around Mayan ruins. Despite its cultural importance, the Oscellated Turkey was never tamed. To this day, zoos cannot coerce the species to breed when held captive, or to lay eggs when penned.

The wild stock of the Common Turkey, *Meleagris gallopavo*, is almost as difficult to breed in captivity. Wariness and aversion to confinement characterize all six geographic races of wild turkeys, which range from the Great Lakes to Veracruz, Mexico. These are gangly, nervous birds, which, when confined, often kill themselves in violent attempts at escape. As zoologist Starker Leopold observed, the wild turkey has not had its torso plumped up for table use; it maintains "a slender, fusiform body better fitted for an active wild bird that must run and fly for its life."

It is astonishing that such a bird was ever tamed, let alone domesticated. Even when hatched in incubators and raised in pens, Eastern wild turkeys seldom stay around human settlements for more than a year before they take flight, never to return again. Zoo keepers and game farm managers have had low success rates in breeding wild turkeys with one another in pens. Decades ago, they resorted to hybridizing wild with domestic turkeys in attempts to keep a "wild-colored bird" penned for more than a limited amount of time.

Enigmatic historic documents suggest that this same Eastern race of wild turkeys, while never genetically domesticated, was sometimes kept by Indians of the temperate deciduous forests. Young poults, captured in the wild and kept near Indian huts, were seen by earlier European explorers along Chesapeake Bay, and in Texas, Arkansas, and Alabama. At times, hunters used these captive turkeys as "decoys" to attract free-ranging turkeys.

While Woodland Indians sometimes kept a few turkeys before butchering them, these birds probably did not breed in captivity. Moreover, wild turkeys were abundant in the East, and were among the game most frequently consumed by prehistoric Indians there. Zoological historian A. W. Schorger has estimated that the pre-Columbian population of wild turkeys in the present-day United States was on the order of ten million birds. Hundreds of birds might be found within a half day's hunt from a village, particularly in the rich bottomland forests of the Mississippi watershed. Through the 1880's, pioneers in the Indian Territory frequently spoke of killing a hundred turkeys on a single hunt. Under aboriginal conditions in the East and Midwest, there was little need to keep turkeys at home, for an abundance was within easy reach. Why go to the trouble to domesticate a species when you live in the thick of it?

Where, then, and under what conditions could prehistoric Indians have found it advantageous to trick the wary turkey into breeding within captivity? The place and period of turkey domestication remains unknown, although a Mexican race of wild turkeys was the probable progenitor of all tamed breeds. Turkey bones appear at cave sites in the Tehuacán Valley of Mexico that are 7000 to 9000 years old, but it is unlikely that these are of domesticated birds. Not until 2400 to 2050 years ago were turkeys a significant resource in Tehuacán's hamlets, towns, and villages. Perhaps as the human population increased and altered turkey habitats by burning, farming, and fuelwood cutting, the birds were encouraged to feed and breed around households and farmsteads where food and shelter were assured.

Somehow the fowl's wariness of cleared woodlands and human beings was suspended. Starker Leopold sensed that this must have been a "gradual selective process by which the genetic constitution of the wild bird was modified to bring about a physiological adaptation to a symbiotic existence with man."

It is not at all clear that the first genetically altered Mexican turkeys appeared in Mesoamerican villages. The earliest archaeological evidence of a small Indian domestic breed is from Tularosa Cave in southern New Mexico. There, 2200-year-old bones of four turkeys were found in cultural remains that predated the introduction of pottery. Did this small domestic turkey breed develop in the dry lands north of Mesoamerica, perhaps in the semi-arid uplands of northwest Mexico or the U.S. Southwest?

Dr. Amadeo Rea has contemplated this latter possibility while studying prehistoric turkey bones from the Southwest. After the extinction of *M. crassipes*, this region may have lacked any wild turkey for hundreds if not thousands of years. Through trade with the Mesoamerican cultures to the south, Southwesterners undoubtedly encountered this useful bird, maybe already in a semi-tamed state. By breeding it in an area where it was isolated from wild populations, prehistoric Southwesterners may have accelerated the rate of its domestication and divergence from the wild types.

In any case, turkeys and macaws became spiritual symbols in Southwestern pueblos, which suggests that the pueblos absorbed some of the culture and religion of Mesoamerica. Scarlet macaws came to symbolize Quetzalcoatl in Mesoamerican-derived ceremonies. They became important trade items from the south, because they could not breed north of present-day Mexico. Charmion McKusick has noted that in Southwestern ruins, "where macaws are found in number, turkey sacrifices in some form are also found." She believes that turkey shanks symbolized one form of Tezcatlipoca, the deity who counterbalanced Quetzalcoatl in Mesoamerican cosmology.

When Tezcatlipoca first appeared in Mesoamerican art nine or ten centuries ago, he came, evidently, as an import from the north via the Chichimecans. Traders linked the U.S. Southwest and the Mexican Northwest with Central Mexico for centuries, and were no doubt involved in carrying the turkeys and scarlet macaws back and forth. At later dates, domestic turkeys, bred at Point of the Pines in present-day Arizona, appear at the prehistoric trade center of Casas Grandes, Chihuahua, where they may have been exchanged for macaws, which could not breed in the cold of the north.

The earliest use of turkeys in the Southwest was for making winter clothing. Warm, lightweight robes were fashioned of turkey feather cordage in areas too cool for much cotton production. Prior to the Basketmaker III period, rabbitskin robes were worn during the winter months, but these were quickly and widely replaced by the weavings of turkey feathers and vegetal cordage. In fact, turkey robes surface in ruins located beyond where turkeys themselves are known to have been raised, perhaps owing to trade.

It is hard for some modern-day Americans to imagine that some of their predecessors on this continent valued turkeys more highly for their feathers than for their meat. This preference persisted well into historic times, and was commented on by the conquistadors and padres who first visited Southwestern pueblos.

On the first formal Spanish expedition into the Upper Southwest in 1540, Francisco Vásquez de Coronado was told that the Zuni did not eat turkeys at all, but kept them for their feathers. The conquistador simply did not believe them. He observed that the Zuni kept just a few turkeys, but they were larger than those that he had seen in Mexico. Coronado's party was struck by the feathered quilts used by the Zuni, and by the plumes or feathered prayersticks offered to water spirits left at their springs.

By Pueblo II times, hacked bones left around habitations suggest turkey butchering for meat consumption. Perhaps then, as today in highland Guatemala, turkeys were sacrificed at the time of sowing crops to sustain the planters, for at that time of the year the grain and bean reserves would have already been depleted.

The bones themselves were often saved for making beads, flutes, awls, and whistles. Turkey eggs were used as ritual offerings on the Rio Grande. Their yolks were used to glaze masks worn in Hopi ceremonies. McKusick and others have even suggested that turkeys may have been kept to control grasshoppers and other pests that would otherwise damage Puebloan field crops. Like guinea fowl, turkeys can be effective in reducing grasshopper populations. McKusick believes that in Pueblo III times, as populations increased in Anasazi settlements and fields became more numerous, the role of turkeys in crop protection became more important.

Still, the primary selection pressure placed on turkeys by prehistoric breeders was for plumage variety. Turkeys may have been good to eat, and good for eating pests, but they were also good-looking. Early Spanish explorers in the Americas encountered mixed flocks of red, white, brown, and black birds, with gobblers that displayed brilliant blues, purples, and reds on their necks. The live birds, and the feathered products derived from them, were selected to be pleasing to the eye.

Geneticists this century have discovered that selection for variation in feather color does not necessarily reduce a bird's wild survivability. The heritable traits for wildness and those for plumage are not strongly linked. Despite all the selection for aberrant plumage that occurred in the prehistoric Southwest, at least one of the breeds was still able to get loose from human hands, and to survive and persist in the wild.

We may be seeing the remnants of this regressive breed above 6000 feet along the Mogollon Rim, a scarp that zigzags diagonally across the U.S. Southwest. Prehistoric ruins from there contain turkey bones that are slightly longer and thinner than those of the Indian breeds, showing signs that the animals had been hunted rather than confined. Zooarchaeologists McKusick and Rea believe that these may be signs that the Larger Indian Domestic breed headed for the hills sometime after its introduction to New Mexico and eastern Arizona. They hypothesize that it became the Merriam's wild turkey that now inhabits much of the Southwestern uplands, thanks to its tenacity as well as to several modern reintroductions. The word "feral" rather than "wild" might better describe Merriam's bird. Like the mustang, it may be derived from domestic stock, but it has retained or even gained enough wild behavior and hardiness to thrive in the rugged mountains of the Southwest. Since Southwestern Indians stopped tending the indigenous domestic breeds a century or so ago, the Merriam's wild turkey is the closest surviving kin of the bird that contributed the entire poultry legacy of prehistoric North America.

It is clear that overspecialization, to the point of obligate dependency on humankind, did not occur with the Large Indian domestic breed in the Southwest. However large it grew, it was still not overburdened with meat as are the modern breeds. Indians in the hinterlands diversified their birds' skin and feather coloration, their diets, and perhaps their environmental tolerances, but there is little evidence that they selected for meat production. Instead, turkeys were regarded as sources of multiple products and as powerful symbols.

Because turkeys were valued in so many ways, was it considered foolhardy to breed forms that offered only one superior feature at the expense of the rest? Prehistoric turkey breeding was a folk science, a relatively slow cultural process compared to that which devised the Beltsville breeds via rapid Mendelian genetic manipulation for "icing-on-the-cake" traits.

In prehistoric times, several generations of turkey breeders may have worked in concert. One person's vision of the ideal turkey was constrained by the notions of those who came before him and the available pool of genes. It was also restrained by the opinions of his or her peers with regard to what was considered an aesthetically pleasing and practical bird. And of course, one generation's breeding work was further tempered by those that followed. The net results had more to do with the cultural values shared by folk breeders through time than with the intellectual capacity of a full-time geneticist acting on his own.

VI.

Emanual Breitburg placed the calipers next to the box of turkey tarsometatarsi he had been analyzing and came out of the huge storage room of archaeological specimens to speak with us. Emanual measures these bones, compares their shapes for three-dimensional features that reflect differences among various turkey breeds and subspecies, and plugs these data into a computer to see how turkeys from various places and times cluster statistically. A Southern Illinois University graduate student in physical anthropology, Emanual was traveling through the Southwest and Mexico to increase his number of samples of bones from Indian breeds and their wild relatives.

Amadeo Rea and I met Emanual in a National Park Service conference room, where we pored over his dendrograms and scatter graphs. He had roughed out these preliminary assessments in attempts to discover relationships between various turkey species, races, and breeds.

"Plug in the date of the bones, their latitude, longitude, and elevation, and you get a map of the Southwest," Emanual explained. "You can even see in these relationships possible trade routes that carried Indian turkeys." He showed us how closely the mixed turkey populations from Casas Grandes, Chihuahua, Mexico, and Point of the Pines, Arizona, fit with one another—as if the turkeys' bones had simply been carried a few miles down the road from one place to the other. And yet the distance between these ruins is some 250 to 300 miles.

Emanual then pulled out another graph, this one charting the morphological relationships between all the turkeys he had measured so far from North and Central America. I glanced down at the constellation of dots floating on the paper, and noticed one, off in space by itself.

"What's that?" I asked.

"Oh, those are the bones of turkeys from a modern market on the East Coast," Emanual explained. "Look how all the wild races and Indian domestics cluster together tightly, compared to the market bird. It might as well be an altogether different animal . . ."

In bone thickness, body shape, and behavior, the modern domestic turkey was indeed a different bird by the time it reached the East Coast a few centuries ago. According to Schorger, tamed turkeys from Central America reached Spain between 1498 and 1511; and by 1541, they had arrived in England, where they became exceedingly popular. In 1607, when domestic turkeys from England arrived at Jamestown, Virginia, they were undoubtedly easy to distinguish from Eastern wild turkeys. Catesby, a natural historian of the Southeast, commented in 1731 on the pronounced differences in shape, stature, and beauty between the two birds. He preferred the wild ones because of their adaptability and their tastiness of flesh.

Modern domestic turkeys can grow up to 22 pounds in as many weeks. The selection of rapidly-maturing birds may have begun some time before turkeys were first taken to Europe. Although we lack a good description of those transoceanic travelers, we do know that in its last days the urbanized Aztec empire demanded a constant supply of turkeys from peasants falling under its dominion.

The Aztec poet-king of Texcoco in 1430, Netzahualcoyotl, is said to have exacted a tribute of at least one hundred turkeys daily. Later, Moctezuma required one turkey per villager every twenty days. Some were fed to Moctezuma's menagerie of raptors, some were cooked with corn breads, and others were ritually sacrificed. To meet the demands of the imperialists who ruled them, Mesoamerican countrymen must have had to select more and more for quick-maturing, lushly-colored birds. Were the selection pressures on turkeys in Mesoamerican villages different from those found earlier in the semi-arid hinterlands? I believe so, for in many ways the latter days of the central Mexican civilization were more like those in feudal Europe than those in the dry Southwest. In Europe, the production of both humans and their livestock was appropriated to support the excesses of their overlords.

When they first arrived in Europe, turkeys became the consuming passion of kings, bishops, magistrates, and admirals—an exotic curiosity, or an ideal dish for a royal wedding feast. By the mid-sixteenth century, the progeny of just a few introductions from Mexico were being produced in large numbers, even by peasants. A Spanish cardinal wrote, "The turkey is a bird which has increased wonderfully in a short time. It has been a very good asset, people driving them from Languedoc to Spain in flocks like sheep."

Two and a half centuries after their return from Europe, American turkeys came under another kind of pressure—the beginning of poultry shows' obsession with genetic uniformity. In 1849, the first major American exhibition of domestic fowl attracted more than a thousand birds and ten thousand people to Boston's Public Garden. By 1873, the entrepreneurs involved in such shows decided to standardize the varieties of fowl to certain types, excluding any birds whose morphological features or color patterns deviated from their conception of the breed. The American Poultry Association's manual, *The Standard of Perfection*, contains descriptions of five thoroughbred turkeys, and recommendations for disqualifying from shows any entries that departed from these descriptions. Historians Page Smith and Charles Daniel have suggested that *The Standard of Perfection* can be used as a measuring rod for the decline of true poultry diversity at the expense of superficially ornamental variety.

Seventy years after their first shot at defining acceptable turkeys, the Association had added only two more turkey breeds to their exhibitions. Their reasons for disqualifying minor deviants from their ideotypes grew more numerous. For all intents and purposes, the Association's actions served to promote a kind of elitist eugenics for galliforms—fowl that expressed superficial differences from the norm were removed from the breeding gene pool. If a breeder-exhibitor had a turkey that did not look like one of the seven acceptable shape and color types, he would cull it from his breeding stock, or drown any of its progeny.

As a result, today these birds—lacking the survival skills of their ancestors—are in danger of drowning themselves. These purebred birds simply don't have the survival instinct that their ancestors had. When Starker Leopold was studying the differences between wild and modern domestic turkeys, he suggested that domesticated stock have their senses dulled, or send weaker danger signals to their brains. They seldom show the "freeze or flight" fear reactions that wild turkeys do. Instead they exhibit maladaptive or lackadaisical behavior in response to dangers that wild birds do not allow to threaten their survival.

At the same time, Leopold showed that the domesticated fowl have a *lower threshold of stimulation* when it comes to sexual excitement and breeding. They attain sexual maturity earlier, and attempt to breed at times of the year that would be inappropriate for survival in natural habitats. Wild turkeys must meet more exacting conditions in terms of season, weather, and habitat before they will initiate mating, but the survival rates of their chicks are higher than those of domestic or hybrid turkeys released into the same areas. In captivity, however, domestic turkeys can breed earlier without the losses suffered in the wild, and so their lack of discretion is less costly.

Wildness, it seems, has to do with the appropriateness of one's responses to unpredictable conditions. Leopold noted not only that wild turkeys are wary and shy, but that nesting hens and chicks consistently react to danger in ways that are self-protective. For much of the year, they travel and forage in small flocks. They are highly selective of habitat, and do not tolerate the large flock sizes and disturbances tolerated by domestics.

To be so finely tuned to their environment, turkeys must have both central and sympathetic nervous systems that pick up subtle cues and respond appropriately. This is no accident, but the result of many years of natural selection in the wild. The adrenal glands of wild turkeys are 100 percent larger, their pituitaries are 50 percent larger, and their brains are 35 percent heavier than those of domestic breeds. Brains and glands, not just meat, make the modern breed an altogether different bird, and a tragic one. It has lost a significant amount of its wildness, and hence its capacity to survive in anything but the most pampered and unnatural conditions.

VII.

Turkey on the loose. The mixed flocks of turkeys that Amadeo Rea and I had seen ranged out freely from their Mountain Pima Indian homesteads into the fallow fields and pine-oak woodland slopes of the Sierra Madre. At dusk, while we talked to Pima families at their homesteads, we watched the gobblers and hens come in to roost in trees a few paces from the doorstep where the Indians' dogs kept guard. Our curiosity was piqued, for these birds seemed hardy, and diverse in color, size, and shape.

They were still on our minds as we travelled across the Sierra Madre Occidental, leaving behind the pine forests of Chihuahua, to begin a descent that would eventually take us to the coastal Sonoran Desert, a land too hot for turkeys. There was something about those birds that Amadeo could not quite put his finger on, and he hoped that there would be someone to ask about them when we reached the last Pima villages lower down. Although the other settlements looked close on the map—just sixty to one hundred miles—it took hours and hours of lunging and plunging our four-wheel-drive vehicle up and down bedrock roads to reach Yecora, Sonora. There we hoped to meet one of the Duarte brothers, who belonged to a Pima farming family with long tenure on an isolated rancho in the heart of the sierra. The Duartes, we were told, knew seeds and breeds as well as anyone.

Leonardo Duarte had left the ranch in recent years, but neither the ranch nor Piman traditions had left him. As he and his wife walked us around their yard—part garden, part menagerie, part materials storage place—they named the plants, products, and creatures in Piman. We asked them about the *tova*, a Piman word for turkeys used over a thousand-mile stretch of sierra and desert. They knew the word, but tended to speak of wild and domestic turkeys as *coconos*, a Hispanicized Indian term in wide use over Mexico. Leonardo responded to our questions with descriptions of the birds he had kept and of the wild ones in the hills nearby. Though we neither taped nor took notes while talking to Leonardo, the insight and subtlety of his comments remain clear in my mind although five years have passed.

"We usually guard a small flock of tame turkeys," he said, "but sometimes in the winter, the wild *coconos* come in to feed with them. Or when we are in the mountains, we see a young one, chase it down, and carry it back in our arms. Those *coconos* will stay through the winter as long as we feed them, which we do, hoping to fatten them up for eating ourselves. But then, around April, after mating with the tame *cocono* hens here, those *machos* fly off, to breed with the wild hens in the mountains."

Leonardo laughed, and added, "They leave their mark with the hens in my houseyard, and back in the wild too."

The best of both lives? We can only vaguely imagine that possibility. Strictly speaking, our own cross-compatible wild relatives are long gone; all that is left of hominid wildness is in our own genes. Lacking kin that are genetically any more wild than ourselves, if we are to be reinvigorated by wildness, it must be behaviorally and spiritually. In one way or another, we may be like the domestic animal breeds that need some periodic renewal of their wildness to keep them hardy and healthy.

VIII.

Wildness may also keep them beautiful, a term that can't be used with a straight face when speaking of the Broad-Breasted White breed after seeing the iridescent feathers of a wild turkey. After hearing Leonardo's account of interbreeding between wild and domestic turkeys, Amadeo Rea said, "So that may be what I'm seeing in turkey feather headdresses from the Tarahumara Easter ceremonies!"

The Tarahumara just to the south of the mountain Pima have maintained their traditional ritual, dress, and language to a greater extent than their neighbors have. Yet their turkey tending, like their agriculture in the sierra, must be somewhat similar. In the 1930s, Robert Zingg recorded that the Tarahumara ran wild turkeys down and captured them alive, or carried their eggs home to put under domestic turkey hens. Young wild-type birds were carefully raised and fed until they would no longer run or fly away. After deer, these birds were among the most important meat sources of the Tarahumara.

We know that turkeys provide feathers as well as meat to the Tarahumara. During Holy Week, dozens of Tarahumaras wear feathers in Indian versions of the Passion Play, in which the Pharisees place on their heads crowns of turkey feathers, or baskets covered with the same. The feathers in Tarahumara crowns that Amadeo mentioned are as intricately patterned and colored as any that he has observed anywhere. Some of the feathers, it seems to him, suggest wild genes; others, hybrids or domestics.

These feathered coronas dazzle the crowds that assemble during *Semana*

Santa, a time of renewal among the Tarahumara. The Pharisees play only one part in the ritual, yet what a striking one it is. Young men daub their bodies in white clay paint. They carry huge drums, and hoot or holler whenever prompted. Bright red bands stream from their foreheads. Scarlet scarves drape their necks. A mound of showy feathers bristles and glimmers as the wind and sun play upon their colors.

It is early April in the wildlands of the sierra. The clear, shrill sound of a gobbler resonates across the canyon. The red on his neck and forehead intensifies as he struts. When he moves into the shadows, a spot of white shows on his tail. But as the light strikes him once again, he flames up into a half dozen colors: purple bronze, copper, greenish gold, buff, black, and a metallic red.

Back across the canyon, the drum drones on beneath the music of flutes. A human community is being resurrected. Men, smeared with earth, crowned with turkey feathers, dance wildness back into their lives.

CHAPTER ELEVEN

Harvest Time:

Northern Plains
Agricultural Change

I.

It is harvest weather. The sky at sunrise is pouring mauve and burnt-orange light out over the undulating land. Farmers' windbreaks, rising between grain fields and wheatgrass pastures, are stained with these same tones: burnt orange and mauve leaves are showering the ground. Only a sprinkle of rain mixes with this morning's wind, but rolling fog threatens to mask the terrain. The weather is pressing humankind to gather what is still left standing in the fields, for soon all will change.

I rose early to begin the drive from Bismarck and Mandan, northeastward toward the geographic center of North America. But before reaching the continental midpoint, I veered westward from the Missouri River at Lake Sakakawea. Following the waters of the reservoir upstream I arrived at New Town, North Dakota, tribal headquarters of the Hidatsa, Arikara (Ree), and Mandan. Along the Upper Missouri, this time of year was once called "the moon of ripe maize." In weather so uncertain, the Indian farmers of the Missouri floodplain would work tirelessly to pull all the maize and squash out of the fields.

Now their former fields lie beneath yards of water behind Garrison Dam. These fields of native maize were at one time so elegant that they deeply impressed botanist John Bradbury, who had encountered corn before in the eastern United States and in tropical Latin America, but none that could compare with the fields on the Upper Missouri in 1811: "I have not seen, even in the United States, any crop of Indian corn in finer order or better managed than the corn about the three villages." He added, "The women . . . are excellent cultivators."

Having read such reports of the historic richness of Mandan, Hidatsa, and Arikara agriculture, I had always admired them. Now, visiting the honey-colored shortgrassed plains and badlands where these tribes live, I admired the beauty of their country as well. At the same time, I realized that 155,000 acres of the Three Affiliated Tribes' holdings—including nearly all of their Class I and Class II farmlands on the fertile bottomlands—were drowned in the 1950s by the filling of Lake Sakakawea, a flood control and irrigation development that serves relatively few non-Indians downstream. The "Village Indians" were forced to move their farmlands away from the river, onto the shortgrass prairies, mesas, and badlands that their ancestors chose not to cultivate.

I was curious about the effects of the move. Like many other ethnic groups around the world, each of these three cultures has within its oral history the memory of being forced out of other regions before settling in the Upper Missouri Valley. Tradition recalls that the Mandan first crossed the Mississippi River at the Falls of Saint Anthony, and archaeologists set the date of Mandan arrival in Dakota country around 1100 A.D. Perhaps by 1300 A.D., the Archaic Mandan became the first village agriculturists of the Upper Missouri. Around 1650, all the Mandan groups that had previously been scattered along the Missouri floodplain were pushed out of South Dakota and southern North Dakota, to become concentrated around the Heart River, a Missouri tributary. This is where the first European visitors to the area found them farming, roughly two hundred and fifty years after Columbus.

The Arikara and Hidatsa have inhabited this stretch of the Missouri for a shorter period of time, but they too have become "native" to the region in terms of the profound way they have adapted to the environment there. The proto-Arikara moved into the Upper Missouri after 1400, after being driven out of the central, semi-arid plains, perhaps by warfare or by an extended drought.

Hidatsa legend maintains that the tribe emerged into its present existence after climbing up a vine from beneath Devil's Lake in northeastern North Dakota. Archaeologists speculate that the linguistically-related, allied groups pushed their way up from northern Illinois into the Red River Valley around 1500, but did not become the unified Hidatsa tribe within North Dakota lands until about 1700. Other scholars suggest that the proto-Hidatsa were a subtribe of the Crow, which split away from the parent tribe around 1650. Sometime before 1700, they obtained corn and other seed from the Mandans, and "relearned" or adapted their agricultural knowledge to fit to their new homeland. Smallpox epidemics beginning in the 1780s, and later threats from the Sioux, encouraged these three tribes to cooperate with one another whenever times got hard.

I felt disturbed and a bit saddened when I arrived in New Town, for I knew that hard times had hit these tribes often, despite a history of them sharing their resources with various neighbors. Given the severity of disruptions that they had suffered since 1790—smallpox, droughts, land grabs, grasshopper infestations, and inundations—it is amazing that they have made their way through to the "New Town" era at all.

When the earlier towns of Van Hook and Sanish were being evacuated as the waters behind the dam rose in the 1950s, the tribes had a contest to decide what they should name their newly combined, relocated community. One suggestion—turned down in favor of the more upbeat name of New Town—was that the names of the old villages should also be combined, so that the community could be known as "Vanish."

The fact remains that the three tribes have not vanished, despite the thirty years their riverbottom farmlands have now been inundated. I wondered if their seeds and agricultural customs had vanished regardless of the persistence of the people themselves. More than anything else, I wanted to hear what these people had to say about any agrarian traditions that might remain.

What was left of the nineteen maize variants that had been described at their villages by various visitors between 1830 and 1920? Did the women still sow any of the kinds of squash that many American gardeners have grown this century? And what became of their handful of bean varieties, among them the progenitor of the Great Northern, at one time the most popular soup bean in America?

My concerns were not limited to the kinds of seeds that had survived, for good farming depends just as much on the survival of appropriate agricultural practices. Did any families continue to save and select seed stocks, using the sophisticated techniques that their forebears had developed? Did the women still cook foods based on these native plants, and had such ethnic specialties reinforced their sense of cultural identity? Had their ways of farming changed when they moved up to the exposed prairie mesas and plains?

The morning clouds began to dissipate. Sunshine broke through, and I too lightened up, a little amused that I had arrived in an unfamiliar community carrying such a storm of questions in my head. Admittedly, I also carried with me something more valuable than curiosity, something I hoped to share with these people even if I never got around to asking a single question.

I had brought along some seeds, long ago collected from these tribes, seeds that deserved to grow in this watershed again, among these people. I had small red Hidatsa beans, glossy and lovely in the palm of my hand. And Arikara winter squash, an early Hubbard type, with blue-green stripes on salmon skin, and puffy, pale, oval seeds. These seeds were given to me by heirloom seedsaver Glenn Drowns. I also carried seeds of a Mandan yellow pumpkin, last grown by Dan Zwiener, and similar in seed size and shape to the early *Cucurbita pepo* gourds raised in the Mississippi watershed thousands of years ago. And the fine-tasting Arikara "yellow" beans, long and kidney-shaped, ranging in color from beige to yellow-orange.

These seeds had been donated by members of the Seed Savers Exchange, but were progeny of seeds collected at Fort Berthold almost a century ago. The varieties I had with me are mentioned in Oscar H. Will & Co. seed catalogs at least as far back as 1913, but that seedhouse continued to feature them until the late 1950s. Sometime along the way, they had gotten into the hands of an heirloom seed collector intent on seeing them survive even if they were never portrayed in a commercial catalog again. And so, they had been passed, season to season, hand to hand, until they had landed in my luggage, bound for their homeland once more.

II.

The flaw in my plan was that I needed to find gardens in order to find gardeners, and in New Town there were few to be seen. Gardens were not common in the other large communities on the reservation, either. On the drive in toward New Town, I had cruised White Shield on the eastern Arikara side of the reservation to ask about a family that had once farmed quite a bit.

One of the men of this White Shield family talked with me at their newer tract house, miles away from where their fields once were. His response was polite, perfunctory, and the same that I heard elsewhere the next two days.

"No, we don't garden anymore," he said in an oddly cheerful tone. "Since we moved from out there in the country, we don't grow any corn anymore. It's been years since we moved away from it all."

I searched for gardens and fields of mixed crops as I passed homesteads that still remained "out in the country," but the few that I did see were of non-Indian families. About 45 percent of the remaining agricultural lands on the Fort Berthold Reservation has either been owned or leased by whites, who now plow large tracts of the shortgrass prairie for barley, wheat, oilseed sunflower, and safflower. The gardens they keep are filled mostly with zucchinis, hybrid sweet corn, and short-season tomatoes. In contrast, members of the three tribes have never become accustomed to tilling the upland soils in their new backyards for similar fields or gardens. Perceived as major limiting factors are the marauding gangs of dogs, children, and urban birds that might do considerable damage. Women repeatedly told me, "You just can't garden in town."

I remembered that 69 percent of the families had successful gardens back in 1948 when they were still in the bottomlands. And every year, the same women who tended those gardens with the help of their children canned or dried 23,000 quarts of wild fruit, gathered mostly from the floodplain and the wooded draws beneath the mesa tops. Wild plums, juneberries, choke-cherries, and a ground bean were plants that proliferated along river banks or moist slopes above side streams.

I soon gained a sense of where the remaining "country people" lived, and how to tell a non-Indian lessee's farm operation or in-town house from those of tribal members. After roaming a while, I came upon the gardens of the Lone Fight brothers and sisters. They clustered around the old country house of their aunt, Mary White Body, on the edge of New Sanish, over-looking the lake.

A couple of frolicking boys chased each other around, through the three small gardens, and over to where my car was parked. "My aunt—she's not home. We'll go get my mother." The boys scurried indoors.

When they bolted back out of the door, soon followed by their mother, their cheeks were full, and they were chewing on a sugary brown chunk of some homemade sweet.

"Cornballs!" I exclaimed to myself, thinking of traveler Henry the Younger's first taste of them in 1806. He was presented with "a dish containing several balls, about the size of a hen's egg, made of pears [juneberries], dried meat, and parched corn, beaten together in a mortar. . . . Boiled for a short time . . . we found them most wholesome."

"Slow down on those cornballs, you boys!" Donna Lone Fight sighed. "Save some for later on!" After introductions, Donna and I wandered over to a garden planted by her twin brother, Donald. There, beets, watermelons, sweet corn, carrots, and honeydew plants were coming to the end of their season, but had produced amply. Donna pointed across the grass to a blue corn crop another brother had planted, and in a different direction to a third plot. I felt some relief—at least there was one native crop maturing in the garden.

"Did your family plant more before?" I asked.

"They did, but it all came to a stop around 1952 or 1953. They were from Old Sanish, and it's all underwater now. Trees, fields, schoolhouse, home. They say you can see a lot of it, standing there beneath the water. So many of them moved to New Town then . . ." They haven't grown any of the old things over there, she explained. As another community member said when he saw the kinds of seeds I gave Donna, "There is so much that we forgot to take with us before the flood."

The boys ran past, still taking bites out of their cornballs. Donna had them run and get me one.

"That's an old kind of food here. Not many people make it anymore. An Indian woman over in Parshall comes around selling them now and then. Maybe she grows a lot of corn, or knows someone who does. She must use an awful lot of it."

I thanked Donna and the kids, gave them some seeds to try for the following year, and headed over to Parshall, fifteen miles away.

III.

Vera Bracklin sat on a living room chair, exhausted, holding her granddaughter in her lap. She had just fed ten families who were mourning loved ones who had gone on.

No doubt Vera was used to large-scale cooking. Sometimes in one batch she would cook up thirty quarts of flour corn, juneberries or blueberries, kidney tallow, and sugar into 120 to 140 cornballs.

"What we call it . . . in Hidatsa, is *mah-pi*. For cornballs, you need soft corn, flour corn, to make it. Mine is corn from my mother-in-law. I can't raise it here, but there's a white lady who gives me room to plant out at her place." Vera Bracklin processed as much corn as she could get her hands on, for the door-to-door sales of cornballs helped to keep her family afloat.

It was the kind of work women in her family had done for a long time. She remembered the effort her grandmother made to cleanly thresh all their Great Northern Beans: "My grandmother used to raise a lot of beans. On a windy day, she'd put a tarp down. She'd take a panful of uncleaned beans, and let the wind blow the leaves away."

Vera glanced down, and looked tired again. "But then we lost everything," she recalled. "It seems like they lost the Indian way of living when the dam forced their relocation."

IV.

Vera's sentiments echoed those of Austin Engel immediately after the reservoir had flooded out the villages. The tribes had been paid a sum of five million dollars, to relocate 90 percent of their people, rebuild houses and roads, and develop new lands. And yet, village communities became dispersed in a way that made Engel feel that their "traditional source of stability is gone."

"Farming became more difficult," Engel explained. "We were far from the neighbors with whom we used to exchange work. The big farmer was taking over the West, and we didn't know how to compete, except to lease our land to him."

I recalled that earlier, others too had feared that the native ways might end, particularly after the smallpox epidemic of 1837. There had been epidemics prior to 1804, when Lewis and Clark wintered among these people, but the tribes had recovered in the years that followed. By the 1830s, when such notables as German Prince Alexander Philip Maximilian, Carl Bodmer, and George Catlin stayed with these tribes and recorded their customs, they were prosperous farmers, hunters, gatherers, and traders.

Prince Maximilian noted that they "cultivate . . . without ever manuring the ground, but their fields are on the low banks of the river . . . where the soil is particularly fruitful. . . . They have extremely fine maize of different species."

He was so impressed by their maize, beans, gourds, sunflowers, and tobacco, that he took seeds of these crops back to botanical gardens in Europe, where they flowered and presumably set seed. The transatlantic introduction of New World crops such as these did much to enrich Europe's royal gardens and peasant fields.

Catlin, with his eye for the landscape as a whole, gives us the feel of the extent and density of crops around their settlements: "We trudged back to the little village of earth covered lodges, which were hemmed in, and almost obscured to the eye, by the fields of corn and luxuriant growth of wild sunflowers, and other vegetable productions of the soil."

The Mandan tribal population had grown up to sixteen hundred by 1837. Then smallpox hit them like a tidal wave. A total of fifteen thousand people, native and immigrant, were killed by the disease on the Upper Missouri that year. Only one hundred and fifty Mandans survived, less than a tenth of their population. The Hidatsas were reduced to five hundred.

When John James Audubon came through their villages in 1843, the survivors' lives were in ruins. They were so distraught and weakened that their usually orderly villages had fallen into disrepair, and the stench of garbage and rotting animal remains was everywhere. Low mounds of dirt, under which smallpox victims were buried, dominated the surroundings instead of the fields of corn that Catlin had seen. After five or six years, these mounds were still barren of all vegetation. Audubon's travelling companion Edward Harris wrote that "the mortality was too great for them to give the usual burial rites of their people by elevating the bodies on a scaffold as described by Catlin."

Overall, Audubon could hardly believe that these were the same places that Catlin had portrayed as active ceremonial centers, with lovely earthen lodges laid out in regular rows. He wrote that "the sights daily seen will not bear recording; they have dispelled all the romance of Indian life I ever had."

The remnants of the two Sioux-speaking tribes, the Mandan and Hidatsa, abandoned their disease-torn villages to move together to a new location in 1845. There, they began to build another life at Like-A-Fishhook, a bend in the Missouri, where Fort Berthold soon became established. The Arikara families joined them in 1862, and considerable intertribal marriage began, perhaps out of necessity. Nevertheless, the Arikara tongue, a Caddoan—not a Siouxan—language, has persisted to this day, in the midst of the numerically dominant Mandan and Hidatsa.

Their agriculture was not intensively studied until a half century later, yet their ancient folk sciences of bottomland cultivation and seed selection had remained intact. Horticulturist-anthropologist George Will, Sr., wrote in 1930 that "through the terrible catastrophes of the 1830s . . . [and] continual harassing by the Sioux, the three tribes, the Mandan, Arikara, and Hidatsa, preserved their agricultural crops and varieties and carried them down even to the present."

Will's farming ethnographies were complemented by those of ethnobotanist Melvin Gilmore and oral historian Gilbert Wilson. Between the three of them, they documented many aspects of the remarkable folk science that still guided the agriculture of the three tribes after the turn of the century. In some ways, Wilson's recording of the farming knowledge shared by Buffalobird-Woman is the greatest testament to the intelligence of an individual native farmer that we have from this continent.

Buffalobird-Woman, or Maxidiwiac, told Gilbert Wilson how Hidatsa families began to develop new fields on bottomland soils, how they fallowed old ones, and why they chose not to plow up the sod on the shortgrass prairie ground in the hills to farm as the government wanted them to do: "The prairie fields get dry easily and the soil is harder and more difficult to work. Then I think our old way of raising corn is better than the new way taught to us by white men." To prove her point, the seventy-three-year-old woman referred to the quality of maize that she herself had raised in this manner. "Last year, 1911, our agent held an agricultural fair on this reservation. The corn which I sent to the fair took first prize . . . I cultivated the corn exactly as in old times, with a hoe."

Maxidiwiac was aware that "corn could travel," and that strains planted within "travelling distance" of one another could be contaminated. To keep each of them pure, she planted different maize varieties some distance apart from one another, perhaps farther apart than corn pollen would normally travel. This was important, "for varieties had not all the same uses with us." Indiscriminate varietal intermixing would be costly. At the same time, "we Hidatsas knew that slightly different varieties could be produced by planting seeds that varied somewhat from the main stock." Her people selected these variants, not only for color, but for other qualities as well.

In view of their sophisticated practices of seed selection and isolation, is it any wonder that George Will gathered from the three tribes six kinds of flint corns, nine to ten kinds of flour corns, and a sweet corn? These maize variants differed not only in color, but in taste and texture, ear size, number of days to maturity, and bushiness of the foliage.

Most of these strains excelled in their hardiness, for they could survive the harshness of Dakota weather. They were, for the most part, short, heavily suckering plants, with ears developing at or near ground level, within a protective cover of foliage and heavy husks that shielded developing ears from frosts and hailstorms. The seeds could "sprout in spring weather that would rot most varieties of corn," Will claimed, and could produce harvestable ears after "about 60 days in a favorable year . . . rarely more than 70 days."

Such adaptations paid off, not only for the three tribes, but for thousands of white settlers on the Northern Plains who later adopted the native corn and beans. Before Will and others passed the three tribes' native strains on to the Montana Agricultural Experiment Station for evaluation, the European-American farmers in that state hardly grew any corn—they grew less than ten thousand acres in 1909. The dent corns accessible through catalogs from the East and Midwest simply did not have sufficient tolerance of the growing conditions found in the semi-arid West.

The Three Tribes' corns gained regional acceptance following widely publicized experiments by Atkinson and M. L. Wilson at the Montana station, which demonstrated the superiority of the northern flints. Will devoted a decade to promoting them through his family's mail-order seed business and his own writings. By 1924, Montana's corn acreage had increased to 420,000 acres. In fifteen years' time, the three tribes' flint corns had allowed nearly a fiftyfold increase in Montana maize production.

V.

To be sure, these native corn varieties have made a tremendous contribution to Northern Plains agriculture. The fact remained that I had yet to see even a single mature plant of native maize on the Fort Berthold Reservation. And while Vera Bracklin had some corn planted in the garden of a non-Indian friend, she deferred to an older, more knowledgeable woman on the other side of the reservoir. "You should go see Cora Baker near Mandaree," she suggested. The Bakers, she explained, not only grew a variety of crops, but kept other traditions alive as well.

And so, on a rainy Sunday morning, I drove out from Mandaree through the rolling hills that spilled into Bear Den Creek. Fields are fewer on this far western side of the reservation, which gradually climbs into the badlands. The area had once been dismissed as "good country for rattlesnakes and horned toads," but that did not keep some families from the old Lucky Mound village from finding solace there after relocation. Bear Den Creek is perhaps more sinuous and wooded than the now-flooded Lucky Mound Creek, yet it may have been enough like it to have attracted the exiles from the lake floor.

As I came into the Baker homestead from the dirt road that ends at their driveway, I could see their garden. I walked to their door, hearing cornstalks rustling in the wind. Cora, a soft-spoken, intelligent-looking woman with neatly pulled-back gray hair, welcomed me in.

"Some people I've met around here the last few days thought that you would be one who might like to have some of these old seeds," I offered. "They say that you still grow others like these."

"Old seeds?" Cora asked, showing a subdued interest. "Could I see what you mean?"

I pulled out my now-crumpled envelopes from the Seed Savers Exchange, and poured Hidatsa red beans into Cora's hand. She just looked at them, saying nothing, as if seeing an old friend for the first time in years. She sat down, then looked at the other kinds of seed, identifying them and commenting on how they were used.

"May I keep them? May I grow these? Here, put some in envelopes for me. I'll go get you some of our family's Indian corn to try."

Cora came back with two ears of flour corn from her garden and a gallon jar full of seeds from her sister's field. She gave me a small bag filled with seeds from the jar, and also the two ears.

"Keep them separate," she warned me, "because corn can travel." She said that she was trying to sort the blue out of some of her white corn that had become mixed. By recurrent selection and roguing, she was working toward her goal, much as Buffalobird-Woman would have done seventy-five years ago.

And much as George Catlin's companions would have done 150 years ago, during the final harvest prior to the smallpox epidemic.

The continuity was there. Cora's daughter Mary talked with us, expressing the concerns of one who would not let these seeds escape again. "We had so much going for us down at Lucky Mound, it was hard to believe that we could have lost everything."

But everything had not been lost, neither Cora's family's seeds, nor the skills that they passed on to her. She told me how her family separated out the seed corn from that which they would eat, and braided the husks of the seed ears together, in an arc an arm's length in size.

While Mary and Cora spoke about their Hidatsa agricultural customs, I listened to their fine voices and to the falling leaves rustling against the windows and roof. My mind drifted off and settled on George Will's words about the harvest season of 1947: "The season of the year when the Indians of our Northern Plains used to harvest their main crop is here. The soft, hazy autumn days with hot noons and cool nights heralded the full ripening of the crops in the Indian fields. . . . All was bustle and confusion as the women and girls hastened to breakfast early from the always simmering pot of boiled corn, beans and meat which . . . hung over the fire. . . . As the women sat about the pile of corn for husking, the wise old grandmother kept her eyes open for plump, large and straight-rowed ears of pure color. These she took and put aside . . . for braiding."

Listening to Cora Baker speak, and hearing the wind scattering the leaves outside, I knew it was the time for braiding seedstocks together again.

CHAPTER TWELVE

Turning Foxholes into Compost Heaps, Shooting Ranges into Shelterbelts

I.

Shards of blue glass on barren ground. Shattered clay pigeons shot over forty years ago. Rusting K ration cans and shell cases, protruding from the surface of the scarred earth.

I reach down amid this rubble where I am digging a hole to plant a desert-hardy tree. Dog tags have caught my eye. Military IDs on a broken chain: they announce through the rust that decades ago, an Italian guy from Brooklyn prepared for war on this spot.

Yet it is not some combat zone in the dunes of North Africa where I am standing, where Rommels or Hannibals become lauded heroes or hated enemies. I am helping to plant a shelterbelt around a three-acre patch of degraded land, now seeded with native crops, in the midst of the metropolitan Phoenix area. Boasting the fifth-largest urban area in the United States, Phoenicians have cleared and paved or built upon the same amount of land covered by Mexico City. However, Mexico City grew into its current nightmare over the centuries; Phoenix is a monster that has gotten loose only since World War II.

Phoenicians love to speak of their city as one where life rises out of the ashes. As I lower the root ball of an ironwood sapling into the ground, I wonder if the sapling can even endure such desolation. Charcoal, gunpowder, and motor oil discolor the earth nearby. Car exhaust burns my eyes. Bulldozers and jackhammers across the street pound my ears. Two million more people will be added to this valley over the next twenty-five years, bulldozing and burning down more citrus groves and saguaro cactus forests, adding to the ashes.

I stand up and shovel more compost in around the ironwood's roots. I close my eyes, wiping the sweat off my brow. I remember seeing old ironwoods and mesquites standing in the middle of remnant O'odham fields, providing shade for native farmers during their midday rest. These legume trees also rimmed more ancient fields, ones that required dense hedges to keep out large animals. An Indian farmer told me that one particular line of ironwood fenceposts had been in place since he was a boy, fifty years before. Another elderly Indian showed me how he brought the nutrient-rich mulch from beneath these trees into his garden, to enrich the soil there.

I open my eyes. Around me, I see broken bottles, decomposing batteries, cracked potshards, and chipped grindstones. The archaeology of peoples who have left this land.

The vegetative cover is sparse, even meager compared to that of healthy desert plant communities. For a brief time after the turn of the century, a temporary Indian town was set up here. Displaced Maricopa Indians grazed sheep for a few years on the modest vegetation, and then were forced to move on. Later, this patch of land was blasted, gouged, scarred, and run over by the Army, which used it as a munitions dump and artillery range from 1942 to 1946.

Ironically, between the overgrazing and the military maneuvers, this land was part of a two-thousand-acre tract hailed as a natural wonder. In 1914, President Woodrow Wilson declared it Papago Cactus National Monument, a title it held until the year of the Stock Market Crash, 1929. Delisted by the Park Service then, it became known as Papago Park. First managed unwillingly by Arizona Game and Fish scientists who constructed hatcheries nearby, it was later sold to the city of Phoenix, which has developed much of it for golfers and picnickers. For the last half century, the Desert Botanical Garden has managed one hundred and fifty acres of the Park to feature desert succulents, cacti, and shrubs, but the military continued to have access to much of this acreage during the war years.

The Botanical Garden has done much to diversify the plant and bird life along trails of horticultural plantings several hundred yards to the south of the old munitions dump, but the flora of the dump itself has become more and more impoverished over the years. Wildflowers that persisted even into the forties cannot be found in the Garden's wildlands today, their seed reservoirs having been gradually depleted in the soil. Several species of cacti, recorded among historic packrat middens in rock crevices nearby, are now nowhere to be found in the remaining "natural" vegetation. Introduced Mediterranean weeds outcompete native herbs for the little moisture that sporadically puddles up in depressions. And, there may have been a loss of certain soil microorganisms such as fungal mycorrhizae and rhizobial bacteria, which normally help many desert plants gain nutrients, thereby aiding in their establishment. Much of the ground surface stays bare the year round.

It is late spring, and over a hundred degrees Fahrenheit when a few of the Garden's staff and volunteers help me to plant the beginnings of the shelterbelt. We hope to buffer our desert food plants from the adjacent traffic's heavy metals and jarring noise. We transplant a dozen species of trees that can produce nutritional, medicinal, or textile products, plus shade and organic matter for plants growing in their shadows. A U-shaped ridge is placed around each one, with a shallow basin dug in front of each tree. The open sides of these basins face upstream, so that if it rains, they will fill with runoff and retain this supplemental moisture. An ugly gouge of an arroyo, which runs through this cultivated patch, is blocked with gabillon check dams that will heal the channel downcutting and funnel storm runoff toward the trees.

But it hardly rains all spring. Summer solstice comes and goes, and we must water the trees by hose, for the thunderstorms will not arrive until late July. Dozens of days of hundred-degree temperatures deplete the trees of their moisture reserves. A couple of the trees die after being damaged by rabbits and rodents, which have little else to eat. Other trees drop their leaves during the worst of the drought, and waver on the edge of desiccation.

I walk among the trees, periodically giving them just enough water to survive. Not fifty yards from the shelterbelt, an artificial river called the Crosscut Canal flows by, tons of water evaporating from its surface every day. Much of the canal water goes to flood irrigate lawns of introduced turfgrasses. Anxious, I wait and wait for rain. No other food crops, not even hardy natives, should be planted until then.

II.

When the rain comes, on July 31st, it is all at once. We have gone over two months with no more than an occasional sprinkle, so light that the dry earth looked like talcum powder again within a few hours following the rain. This time, however, one and eight hundredths of an inch falls within two hours. The arroyo fills to the brim, running waist high, cascading over the check dams. Flood waters are diverted onto the tree plantings, and within two days, many have sprouted new leaves.

The downpour arrives just as we have finished transplanting thousands of tepary bean and millet seedlings into an intercropping experiment. I have gone back on my vow not to plant the food crops until the first good downpour came, but the storms begin within three hours after the last seedling has been placed in the earth.

By this time, the number of people working with these native crops has grown. For the prior two years, the farm plot has been cared for only by Dr. Howard Scott Gentry, an economic botanist in his eighties, and by his part-time assistant. When Dr. Gentry retired in the spring, horticulturist Suzanne Nelson and I took over the desert crop research that he had begun, expanding it to include more species and other planting designs.

In midsummer, Culver Cassa, a nineteen-year-old Pima Indian fascinated with desert plants, joined us as the Garden's Native American intern.

Dr. Cay Randall, an Arizona-born entomologist, soon began looking at the insects that had hatched with the rains, and at others which had migrated in, attracted to our crops. Dr. Rob Robichaux, a plant physiologist who designed the intercropping experiments, helped with the planting and guided our data-taking. Rob brought along his students from Ethiopia, the Philippines, and California. When Suzanne joined Rob for further agroecology studies at the University of Arizona, Mark Slater came out of the Chiricahua Mountains to assist with the daily plant care and data gathering. Mark had done similar work on sustainable agriculture with Wes Jackson at the Land Institute. Soon six scarecrows, decked in blue jeans, rumpled hats, and raggedy plaid shirts, were erected to serve as our vigilantes.

It is a group nearly as heterogeneous as the plants themselves. By summer's end, we are evaluating or experimenting with fifty useful plants native to the Sonoran and Chihuahuan Deserts. We have roughly tripled the number of desert plant species that are growing on that degraded plot of land. Our runoff catchment basins, arroyo diversions for flood irrigation, and drip systems have multiplied, as we work to conserve water. Our goal of a rainfed, desert plant community restoration seems more and more plausible.

Finally, we turn an old army foxhole into a compost pit. We run our crop trimmings and weeds through a chipper-shredder, and they fly onto the decomposing pile that fills the foxhole. With pitchfork and shovel, we turn the pile over, spray it down, cover it with plastic, and let it steam. A transformation has begun, and soon we will have more organic matter to add to the impoverished desert soil.

III.

In early autumn, we hold a training workshop for Mexicans and Native Americans on intercropping and forestry strategies for conserving diversity in desert agriculture. Two dozen scientists, farmers, gardeners, and rural development workers participate. With these friends and teachers among us, we rechristen the farm the Gentry Agroecology Project to honor its founder, Howard Scott Gentry. It is also the wedding of desert ecology and native agriculture that we celebrate. Near the GAP farm gate, we erect a sign with Dr. Gentry's words, reading, "Caution: Plants at Work." We hope to make better use of the ecological adaptations that allow certain native crops to function in extreme heat and drought. Tepary beans. Sonoran panicgrass. Murphey's mescal. Land races of prickly pears from the Chihuahuan Desert. The annual crops have produced plentiful seeds in our small field trials this season, and the perennials have grown well, despite above-average heat and little rain. Some observers tell us that by further breeding these "specialty" crops, we could enable desert farmers to profit on the hundreds of thousands of water-scarce acres where they are now failing.

Even so, we will have missed the point if we only select one of these profitable natives, create new hybrids with it, and grow them in monocultures just like any conventional cash crop. The Native American agricultural legacy is more than a few hardy, tasty cultigens waiting to be "cleaned up" genetically for consumers, and then commercialized as novelty foods. Our goal must be something beyond blue corn chips, tepary bean party dips, amaranth candy, sunflower seed snacks, and ornamental chiles. These nutritious crops deserve to be revived as mainstays of human diets, and not treated as passing curiosities. These cultivated foods are rich in taste and nutrition, yes, but they are also well adapted to the peculiarities of our land. No one has made this point more succinctly than ethnobotanist Melvin Gilmore, who wrote seventy years ago that: "We shall make the best and most economical use of our land when our population shall become adjusted in habit to the natural conditions. The country cannot be wholly made over and adjusted to a people of foreign habits and tastes. There are large tracts of land in America whose bounty is wasted because the plants that can be grown on them are not acceptable to our people. This is not because the plants are not useful and desirable but because their valuable qualities are not known. . . . The adjustment of American consumption to American conditions of production will bring about greater improvement in conditions of life than any other material agency."

I think about Gilmore's words on the days that I catalog the native species that volunteer in and around our small cultivated plots: wild devil's claw, Coyote's tobacco, and palo verde seedlings. These are the plants that also emerge in the floodplain fields of Sonoran Desert Indians. Often, instead of hoeing them out, desert Indians let them grow. Some function as ground cover or are associated with beneficial insects. Their products are harvested, and stories are told about them. My hoe now spares them as well. I wait to see what they will have to offer. When an herbalist visits us from northern Arizona, she notices the Coyote's tobacco left in our fields. "I know some elders who could use that tobacco in their kiva ceremonies," she tells us. We save the tobacco for them.

In essence, what many Native American farming traditions integrate with wild species within their cultivated fields and domestic economies is a dynamic balance of *wildness* with *culture*. This is what modern farmers lose when they cultivate their fields from edge to edge, leaving no hedges, no weeds, no wildlife habitat. The trend in industrial farming is, in fact, a repudiation of wildness. And yet, a certain wildness may be exactly what our ailing agricultural system needs. As Dr. Jack Harlan has pointed out, "Students of crop evolution have a tendency, it seems, to see plant domestication in a consistently positive way. All is for the good and nothing for the bad. . . . Not so familiar, perhaps, are changes that are less beneficial or negative. . . . [For instance,] the photosynthetic rates of wild wheat are at least double those of cultivated wheat at high light intensities. . . . The wild races of cereals almost always have more protein in their grains than cultivated races. [There are] natural defenses deployed by stands of wild wheat, barley, and oats in the Near East [where] one can encounter the same pathogens and even the same races that occur in North America. [In plants genetically resistant to diseases and pests,] the natural defenses turn out to be far more complex than those we have attempted to deploy. . . . There is a complex mix of strong genes for resistance. . . . We may expect plant breeding strategies in the future to be modeled along the lines of wild populations."

At the Gentry Agroecology Project (GAP) farm, we give as much attention to useful wild plants as to already-domesticated native crops that have retained their hardiness. Such wild species have contributed greatly to the nutritional, medical, and energy supplies of Native Americans over the centuries, and should not be neglected. Some, like the shrubby Sonoran Desert oregano (*Lippia palmeri*), produce their own insect repellents and grow well on just two irrigations a year even in the hottest, driest deserts. Others, especially mesquite, offer a range of benefits: high-fiber foodstuffs, furniture wood, honey, and favorable microclimates for other plants. Mesquite pod and seed gums may help Indian diabetics to control their health problems, for when eaten in sufficient quantities, mesquite foods slow down digestion and reduce radical changes in blood sugar levels. Desert Indians today are extremely prone to diabetes, and yet many partial cures and controls can be found right at the edge of their fields. The wild resources of field margins were formerly used by desert tribes as much as their intentionally-sown crops were used.

It is a short step from appreciating a single wild species to appreciating the mutualistic associations between two or more wild plants. We are attempting intercropping experiments that pair useful plants together that are typically found in association with one another in their natural habitats. Thus, by studying which species have functional ecological relationships in the wild, we have a viable model for new polycultures at the GAP farm.

IV.

Several years ago, I became curious about the "nurse plants" commonly found to shade wild perennial chiles on the desert edge. Even where mesquites or oaks were the dominant tree species, I noticed that the chiltepín bushes hugged the trunks of shrubby hackberries and wolfberries. A conversation with a rural Sonoran woman, a chile aficionado, confirmed my observations. Josefina Duran told me, "In all my life, I have never seen a chiltepín bush growing alone, out in the open. It is always under shrubs." Beneath these shrubs, they found protection from freezes and extreme heat, but there was something more.

About the same time, an elderly Pima Indian farmer in Onavas, Sonora, told my colleague Amadeo Rea that the wild chiles were dispersed by red birds such as cardinals and pyrrhuloxias. During part of the year, these birds apparently need sufficient carotene from reddish fruits to maintain their plumage colors, so they feed on red berries. Such berry specialists will pick up chile fruits now and then, which are about the same size, hue, shape, and brightness of the birds' mainstay berries. It is no wonder that wild chiles frequently become established under wolfberries and hackberries, for they are probably dispersed beneath the canopies of these shrubs when the birds pick their berries. Once they have germinated there, they find a buffered microclimate more suitable for growth than open, barren ground.

During the summer, Mark Slater and I have begun to use this natural model in a modest experiment. We plant one chile under the shade of a hackberry sapling, another under artificial shade, and a third in the open. We have already planted thirteen sets of these three ecological conditions for the chiles. They are watered by drip irrigation supplemented by storm runoff collected in small catchments. On a monthly basis, Mark and I record which are surviving, growing, and thriving. It may take two, three, or even five years to see whether a truly native agriculture can emerge from such tests. But already, there are other indications that when we multicrop desert natives, benefits derived from the presence of some plants are spread throughout the entire plant community.

Dr. Cay Randall has been sweeping, trapping, bagging, and observing the many insects that are found among the GAP farm plantings. In addition to the plant-eating leaf miners, leafhoppers, aphids, and whiteflies that are attracted to our crops, there are several beneficial insects that are predators or parasites on these potential pests. Crab spiders are our most consistent, abundant pest predator, and perhaps do a decent job of keeping certain other insect populations in line. Convergent lady-beetles, leafhopper assassin bugs, praying mantids, green lacewings, and umbrella wasps are among the dozen other parasatoids and predators that Cay has encountered.

We have no immediate way to determine whether the cumulative effects of these beneficial insects are enough to keep a massive outbreak of a single pest from occurring among our crops. But Cay has already observed something that is significant. These insects are shared between the surrounding vegetation and our crops. The useful wild shrubs we have planted also support this insect diversity. The dynamics of insect populations in a single large patch of one crop, with wild vegetation cleared around it, would be very different. We intend to draw on the services of a variety of beneficial insects for pest management in our plantings rather than using chemical controls. To do so, we must learn how to foster and maintain these beneficial insects by meeting their habitat needs in the wild perennial plants we keep near our native annual crops. It may be unlikely that many Indian farmers intentionally manipulated wild plants in and around their fields for biological control of crop pests. Nevertheless, they dispersed their small fields amid natural vegetation, and perhaps reduced the probability of pest buildup as a consequence.

V.

Five months after planting the first few trees on the field margins, I walk through the GAP farm one evening at midnight. It is harvest time, and like the Cocopa Indian farmers of the Colorado River, I feel inclined to keep watch in the field, alert to nighttime predators. The Cocopa used to build a "watcher's bower" of branches as makeshift shelters, bedding down in the heart of their fields. Wandering through the gourd patch, the tepary inter-crops, and chile-hackberry basins, I fail to see any rabbits, so I retire to a shelter somewhat similar to the Cocopa bower.

It is a ramada of rooted, sprouted, thorny ocotillo branches where I lie down for a while. I glance at the wild passion flower vines using the ocotillo as a trellis, then scrambling over the palm fronds that form the ramada roof. In the gaps between some leafy ocotillo uprights, I notice a moondog coalescing in the newly arrived cloud cover. I ponder that vaporous ring circling the moon.

"A moondog." I try to recall what I learned while living in Midwestern farm country years ago. "Does a moondog say that rain is soon on its way? The crops sure could use some more moisture . . ."

Finding that I am talking to myself, I concede that I am tired enough to close my eyes. I dismiss the notion of farmers' forecasts, and try to sleep.

The rain begins later that night, and I leave. It pours off and on for two days, drenching the crops, inundating the catchments basins around them, and filling the arroyo with nutrient-rich silt and water. Although we will need to repair a few basins around our plantings afterwards, the floods are a blessing.

The earth smells fertile again. Wild seeds are germinating in the rainfed desert soils. A healing has begun. We have plenty of work to do.

BIBLIOGRAPHIC
ESSAYS

Prologue: Enduring Seeds—The Sacred Lotus and the Common Bean
For an introduction to the longevity of seeds, see my 1978 essay on the folklore and science of archaeological seeds, and the 1986 review by Roos. With regard to how the loss of adapted seedstocks has made agriculture more genetically vulnerable, the National Research Council's 1972 report remains a classic, although I am currently assisting Garrison Wilkes, Ivan Buddenhagen, and Donald Duvick in a new interpretation/edition. Other works consulted or quoted in the prologue include Adams (1977), Conard and colleagues (1984), Kaplan (1981), Jarvis (1908), Lewandowski (1988), Libby (1954), Ohga (1923 and 1926), Priestly and Posthumus (1982), Ramsbottom (1942), Wester (1983), and contributors to Whealy's 1986 compilation of Seed Savers Exchange correspondence and history.

Chapter One: The Flowering of Diversity
The best overview of biological diversity, its importance and endangerment, is the 1988 anthology edited by E. O. Wilson and Frances Peter; it is an outgrowth of presentations given at the National Forum on BioDiversity held in Washington, D.C., in 1986. It includes eloquent commentaries by scientists, conservationists, poets, and philosophers. Many of the concepts relating to the historic diversification of flowering plants were reviewed by Philip Regal in his 1977 article on the evolution of flowering plant dominance. The role of Pleistocene megafauna in shaping American landscapes, and the possible role Native Americans played in extirpating these large animals are reviewed in the 1986 anthology edited by Paul S. Martin and Richard G. Klein. I discuss the relationship between these extinctions and the initiation of intensive food production in the New World in a technical paper in press, inspired by Alford (1970), and Janzen and Martin (1982). The literature on Florissant fossil beds is interesting reading in and of itself. See Saenger (1982) for a popular account, or Leopold and MacGinitie (1972) for recent scientific ruminations. Other literature cited or interpreted includes Beach (1982), Beck (1976), Bordes (1968), Crosby (1986), Crosby and Raven (1985), Charles Darwin in Francis Darwin and

Seward (1903), Hunter and Aarssen (1988), Kruckeberg (1986), MacGinitie (1953), Martin (1986), Pellmyr and Thien (1986), Lewis and Clark in Sweet (1962), Tifney (1984), Toledo (1987), and Whittington and Dyke (1986).

Chapter Two: Diversity Lost: The Wet and the Dry Tropics
Anyone unacquainted with tropical conservation issues should first partake of the popular books by Norman Myers (1984) and Catherine Caulfield (1985). To counterbalance their preoccupation with the wet tropics, I recommend Daniel Janzen's 1988 essay on the dry tropics in *BioDiversity*. With regard to the Guatemalan deforestation problems, I have relied heavily on the excellent reports by Nations and Komer (1984), Wotowiec and Martinez (1984), and Berganza (1987), supplemented by my own brief field trips there with the Peace Corps. With regard to the transition from the desert to the tropics, I have drawn on Wiseman (1980). To compare genetic resources of wet and dry tropics, I have used data from Esquinas-Alcazar (1988) and from my 1986 essay printed in *Annals of Earth Stewards* and in *Earth First!*. Other articles drawn upon for this chapter include Creech cited in Miller (1973), Givnish cited in May (1975); Janzen (1986), Maslow (1987), Prescott-Allen and Prescott-Allen (1987), Simberloff (1987), and various news reports in *Science* magazine regarding tropical deforestation, especially those of Holden.

Chapter Three: Fields Infused with Wildness
This essay evolved out of my 1987 ethnobiology and conservation article, and several lectures. For the discussion of maize-teosinte introgression, it attempts to find some middle ground between Wilkes (1970), and Doebley (1984); Doebley's unpublished analysis of my corn and teosinte samples from Nabogame indicates the probability of minor gene flow from teosinte to corn, but certainly not any swamping. I have also drawn upon the work of Betancourt and Van Devender (1981), Brown (1985), Diamond (1986), Doebley and colleagues (1984), Ehrenfeld (1986), Emslie (1981), Lumholtz (1902), Nabhan and colleagues (1985), Nabhan (1983), Oldfield and Alcorn (1987), Pennington (1969), Vovides (1981), and Winter (1976).

Chapter Four: Invisible Erosion: The Rise and Fall of Native Farming
This chapter echoes the ornithological treatise set forth by Amadeo Rea in his 1983 classic, *Once a River*. There are few other books which discuss subsistence histories of particular tribes in sufficient detail with regard to ecological and genetic changes; William Cronon's 1983 *Changes in the Land* is another notable exception. Carlson (1982) has explored historic government policies which discouraged Indian farm-

ing, and I have questioned other government stances regarding conservation of Native American plant genetic resources in 1985 and 1987 articles. Hurt's 1987 history of Indian agriculture in North America was not available to me at the time of researching this chapter, but it is nevertheless highly recommended. All of these authors are in agreement that small-scale Indian farming did not die out due to any ecological failure, but because of external economic and social pressures. Other literature cited or interpreted includes Bohrer (1970), Bureau of Indian Affairs (1988), Cabeza de Vaca in Smith (1871), Conard and colleagues (1984), Dobyns (1977), Ford (1981), Doyel (1979 and 1988), Haury (1976), Heiser (1985), Masse (1981), Minckley (1973), Spicer (1962), Smail (1987), Sauer (1977), Yarnell (1977), and Yarnell (in press). I must also acknowledge that *The Ballad of Ira Hayes* written by Peter LaFarge and Johnny Cash goes through my head whenever I think of the Pima Indians losing the Gila River; lyrics are used by permission of CBS/Columbia Records.

Chapter Five: A Spirit Earthly Enough: Locally Adapted Crops
and Persistent Cultures
This chapter is a drastic revision of my infamous 1983 Kokopelli essay; I no longer think that the symbol of Kokopelli as a cross-cultural trader of seeds is as key to crop genetic diversity as is each culture's survival. The best technical papers on crop genetic diversity in cultural contexts are by Altieri and Merrick (1987) and by Oldfield and Alcorn (1987). I have also drawn upon Anderson (in press), Basso (1987), Berry (1981), Betancourt and Van Devender (1981), Brush (1986), Burns (1983), Bell and colleagues (1980), Crosby (1986), Franquemont (1987), Gould (1986), Harlan (1975), Lawton and Wilke (1979), LeBlanc (1985), Nabhan (1979), Nabhan and colleagues (1983), Nations (1984), Spicer (1971 and 1980), Trenbath (1975), Wolfe (1985), and van Willigen (1981).

Chapter Six: New and Old Ways of Saving: Botanical Gardens, Seed Banks, Heritage
Farms, and Biosphere Reserves
An excellent introduction to current grassroots efforts to save rare plants and animals was edited by the U.S. Congress Office of Technology Assessment in 1986, based on four commissioned papers, one of which I prepared with Kevin Dahl through Native Seeds/SEARCH. For this chapter, I have also relied upon discussions by Archibald in Houseal and colleagues (1985), Barton (1978), Dugelby (1987), Elias (1987), Emmart (1940), Euler (1956), Grossman (1985), Hirth (1984), Nabhan and colleagues (1981), Plucknett and colleagues (1983), Posey (1984), Ramírez-Acosta (1986), and Whealy (1986).

Chapter Seven: Wild-Rice: The Endangered, the Sacred, and the Tamed
Although not yet available when I was writing and researching this chapter, Thomas Vennum has recently published what will surely stand as the classic on wild-rice and native peoples. Jim Meeker's extensive literature and field notes on wild-rice were my touchstones instead. For this chapter, I consulted Danziger (1978), Emery in Terrell and colleagues (1978), Great Lakes Indian Fish and Wildlife Commission (1987), Jenks (1900), LaDuke (1987), Levi (1956), May and Lyles (1986), Meeker and Ragland (1987), Nelson and Dahl (1986), Oelke (1974), Ritzenthaler (1978), San Marcos Recovery Team (1984), Taylor in Danziger (1978), Terrell and Wiser (1975), and Thannum (1987).

Chapter Eight: The Exile and the Holy Anomaly: Wild American Sunflowers
This chapter could not have developed without the sunflower research pioneered by Charles Heiser (1985, etc.) and by Rogers and colleagues (1982). Recent reviews of wild sunflower contributions to the economy include Laferriere (1986) and Prescott-Allen and Prescott-Allen (1987). I have also drawn upon Chandler and Beard (1983), Clifton and Callizo (1986), Gathman and Bemis (1983), Jain and colleagues (1977), Kruckeberg (1986), Lewis and Lewis (1981), Nabhan (1983), Nabhan and Reichhardt (1983), Olivieri and Jain (1977), and Rieseberg and colleagues (1988).

Chapter Nine: Lost Gourds and Spent Soils on the Shores of Okeechobee
Marjorie Stoneman Douglas (1978) and the Hannas (1948) have written excellent histories of the Everglades and the Lake. Blake's (1980) water history and Will's (1977) Cracker history also provide insight into this regional environmental tragedy. Two opposing views of the taxonomy of the gourd—Andres and Nabhan (in press) versus Robinson and Pulchalski (1980)—have been used as an excuse by state and federal agencies for their failure to provide protection to this plant, although both sets of authors have repeatedly stated that the plant urgently needs protection. I have interpreted material from Buckingham Smith (1848), Decker (1987), Dobyns (1983), Ehrenfeld (1986), Fernald and Patten (1984), Gifford (1925), Harper (1958), Milanich (1980), Morton (1975), Prescott-Allen and Prescott-Allen (1987), Providenti (1978), and Small (1930).

Chapter Ten: Drowning in a Shallow Gene Pool: The Factory Turkey
This chapter was inspired by Lawrence Alderson's 1978 book on rare breeds, which introduced me to turkey overspecialization. However, Schorger's 1966 masterpiece on turkey history and McKusick's recent monograph on turkey prehistory are the

best starting places for reading on this topic. I have also drawn upon the American Poultry Association (1953), Bennett and Zingg (1935), various authors included in Hewett's 1967 anthology, Leopold (1944), Ligon (1946), Perron (1691), Rea (1980), Schaafsma (1980), Shroeder (1968), and Williams (1981). Special mention must be made of the "fictional excerpt" from Douglas Unger's 1984 *Leaving the Land*, a book with remarkable insight into recent rural American change. The passage is used with permission.

Chapter Eleven: Harvest Time: Northern Plains Agricultural Change

My interest in the crops of the Three Affiliated Tribes has long been invigorated by the writings of Gilmore (1919), Will and Hyde (1917), and Wilson (1917 and 1987). In addition, Meyer's masterful 1977 ethnohistory is critical to understanding historic change. I have also relied upon Audubon in Durant and Harwood (1980), Bradbury (1867), Case (1977), Coues (1897), Macgregor (1949), McDermott (1951), Meyer (1968), Smith (1972), Wilkins and Wilkins (1977), and Will (1947).

Chapter Twelve: Turning Foxholes into Compost Heaps, Shooting Ranges into Shelterbelts

Much of the research alluded to in this chapter is in progress, so that no definitive reports by my colleagues Nelson, Randall, Robichaux, and Slater are available at the time this goes to press. For starters, however, I refer readers to my article with Richard Felger (1985) exploring wild desert relatives of crops. I have also referred to concepts or excerpts from Gilmore (1919), Harlan (1983), Liebman (1987), Trenbath (1975), and Wolfe (1985).

LITERATURE CITED

Adams, M. W. 1977. An estimation of homogeneity in crop plants, with special reference to genetic vulnerability in the dry bean, *Phaseolus vulgaris*. L. Euphytica 26: 665–679.

Alderson, Lawrence. 1978. The Chance to Survive: Rare Breeds in a Changing World. Cameron and Tayleur, London.

Alford, John. 1970. Extinction as a possible factor in the invention of New World agriculture. Professional Geographer 22(3): 120–124.

Altieri, M. A. and L. C. Merrick. 1987. In situ conservation of crop genetic resources through maintenance of traditional farming systems. Economic Botany 41(1): 86–96.

American Poultry Association. 1953. The American Standard of Perfection. American Poultry Association, Atlanta.

Anderson, E. N. In press. Learning from the land otter. Journal of Ethnobiology.

Barton, Frances. 1978. Rare plant sanctuary. The Herbarist 44: 6–14.

Basso, Keith H. 1987. "Stalking with stories": Names, places, and moral narratives among the Western Apache. In: On Nature. Edited by Daniel Halpern. North Point Press, San Francisco. pp. 95–116.

Beach, J. H. 1982. Beetle pollination of *Cyclanthus bipartitus* (Cyclanthaceae). American Journal of Botany 71: 1149–1160.

Beck, Charles B., ed. 1976. Origin and Early Evolution of Angiosperms. Columbia University Press, New York.

Bell, Fillmen, Keith Anderson, and Yvonne G. Stewart. 1980. The Quitobaquito Cemetery and Its History. National Park Service, Western Archaeological Center, Tucson.

Bennett, Wendell C. and Robert M. Zingg. 1935. The Tarahumara: An Indian Tribe of Northern Mexico. University of Chicago Press, Chicago.

Berganza, Gustavo. 1987. CELGUSA: La última decisión. (Domingo) Prensa Libre. 329:8–9 (2 May).

Berry, Wendell. 1981. The Gift of Good Land. North Point Press, San Francisco.

Betancourt, Julio L. and Thomas R. VanDevender. 1981. Holocene vegetation in Chaco Canyon, New Mexico. Science 214: 656–658.

Blake, Nelson Manfred. 1980. Land into Water—Water into Land: A History of Water Management in Florida. University Presses of Florida, Gainesville.

Bohrer, Vorsila L. 1970. Ethnobotanical aspects of Snaketown, a Hohokam Village in Southern Arizona. American Antiquity 35(4): 413–429.

Bordes, François. 1968. Old Stone Age. McGraw-Hill, New York.

Bradbury, John. 1867. Travels in the Interior of America in the Years 1809, 1810 and 1817. In: Early Western Travels. Edited by Ruben Gold Thwaites. Arthur Park, Cleveland. Volume Five.

Brown, Cecil. 1984. Mode of subsistence and folk biological taxonomy. Current Anthropology 26(1): 43–62.

Brown, Cecil H. 1984. Mode of subsistence and folk biological taxonomy. In: Language and Living Things: Uniformities in Folk Classification and Naming. Edited by Cecil Brown. Rutgers University Press, New Brunswick, N. J.

Brush, Stephen B. 1986. Genetic diversity and conservation in traditional farming systems. Journal of Ethnobiology 6(1): 151–167.

Bureau of Indian Affairs. 1988. Indian Agriculture/Range Program. United States Department of Interior, Washington, D. C.

Burns, B. T. 1983. Reconstructing prehistoric crop yields from tree rings, A. D. 650–1970. University of Arizona unpublished dissertation, Tucson.

Carlson, Leonard A. 1982. Indians, Bureaucrats, and Land: The Dawes Act and the Decline of Indian Farming. Greenwood Press, Westport, Connecticut.

Case, Harold W., editor. 1977. One Hundred Years at Fort Berthold. Bismarck Tribune, Bismarck, North Dakota.

Caulfield, Catherine. 1985. In the Rainforest. Alfred Knopf. New York.

Chandler, John M. and Benjamin H. Beard. 1983. Embryo culture of *Helianthus* hybrids. Crop Science 23: 1004–1007.

Clifton, Glen and Joe Callizo. 1986. Monitoring rare plant populations in the Knoxville area of California. In: Conservation and Management of Rare and Endangered Plants. Edited by Thomas S. Elias. California Native Plant Society, Sacramento. pp. 397–400.

Conard, Nicholas, et al. 1984. Accelerator radiocarbon dating of evidence for prehistoric horticulture in Illinois. Nature 308: 443–445.

Coves, Elliot, editor. 1897. New Light on the Early History of the Greater Northwest. The Manuscript Journals of Alexander Henry and of David Thompson, 1799–1814. Ross and Hanes, Minneapolis.

Cronon, William. 1983. Changes in the Land: Indians, Colonists, and the Ecology of New England. Hill and Wang, New York.

Crosby, Alfred W. 1986. Ecological Imperialism: The Biological Expansion of Europe, 900–1900. Cambridge University Press, Cambridge.

Crosby, Marshall R. and Peter H. Raven. 1985. Diversity and Distribution of Wild Plants. Unpublished background paper for the Office of Technology Assessment, Washington, D. C.

Danziger, Jr., Edmond Jefferson. 1978. The Chippewas of Lake Superior. University of Oklahoma Press, Norman.

Darwin, F. and A. C. Seward, eds. 1903. More Letters of Charles Darwin. Appleton, New York. Volume 2.

Decker, Deena Sue. 1987. Numerical analysis of archaeological *Cucurbita pepo* seeds from Hontoon Island, Florida. Society of Ethnobiology Tenth Annual Conference Abstracts, Florida State Museum, Gainesville.

Diamond, Jared. 1986. The environmentalist myth; archaeology. Nature. 324: 19–20.

Dobyns, Henry F. 1977. Who killed the Gila? In: Water in a Thirsty Land. Edited by Henry F. Dobyns. Pinyon Press, Prescott, Arizona. Reprinted from the Journal of Arizona History.

Dobyns, Henry F. 1983. Their Number Become Thinned: Native American Population Dynamics in Eastern North America. University of Tennessee Press, Knoxville.

Doebley, John F. 1984. Maize introgression into teosinte—a reappraisal. Annals of the Missouri Botanical Garden. 71: 1100–1113.

Doebley, John F., Major M. Goodman, and Charles W. Stuber. 1984. Isoenzymatic variation in *Zea* (Graminae). Systematic Botany 9(2): 203–218.

Douglas, Marjory Stoneman. 1978. The Everglades: River of Grass.

Doyel, David E. 1979. The prehistoric Hohokam of the Arizona desert. American Scientist 67: 544–554.

Doyel, David E. 1988. Hohokam cultural dynamics. Paper presented at the Amerind Foundation, Dragoon, Arizona.

Dugelby, Barbara. 1987. Nusagandi Park: a Kuna Indian-run rainforest preserve in Panama. Earth First! June 21: 24–25.

Durant, Mary and Michael Harwood. 1980. On the Road with John James Audubon. Dodd, Mead and Company, New York.

Ehrenfeld, David. 1986. Life in the next millennium: who will be left in earth's community. In: The Last Extinction. Edited by Les Kaufman and Ken Mallory. Massachusetts Institute of Technology Press, Cambridge. pp. 167–186.

Elias, Thomas S. 1987. Can threatened and endangered species be maintained in botanical gardens? In: Conservation and Management of Rare and Endangered Plants. Edited by Thomas S. Elias. California Native Plant Society, Sacramento. pp. 563–566.

Emmart, Emily Walcott. 1940. The Badianus Manuscript: An Aztec Herbal of 1552. The Johns Hopkins University Press, Baltimore. p. 341.

Emslie, Steven D. 1981. Birds and prehistoric agriculture: The New Mexican Pueblos. Human Ecology 9(3): 305–328.

Esquinas-Alcazar, José T. 1988. Plant genetic resources for arid lands: conservation and use. In: Arid Lands Today and Tomorrow. Edited by Emily E. Whitehead and colleagues. Westview Press, Boulder. pp. 397–416.

Euler, Robert C. and Volney H. Jones. 1956. Hermetic sealing as a technique of food preservation among the Indians of the American Southwest. Proceedings of the American Philosophical Society 100(1): 87–98.

Fernald, Edward A. and Donald J. Patten, eds. 1984. Water Resources Atlas of Florida. Florida State University, Tallahassee.

Ford, Richard I. 1981. Gardening and farming before A. D. 1000: patterns of prehistoric cultivation north of Mexico. Journal of Ethnobiology 1 (1): 6–27.

Franquemont, Christine. 1987. Potato breeding in high altitude environments in the Andes. In: Society of Ethnobiology Tenth Annual Conference Abstracts. Florida State Museum, Gainesville.

Gathman, Allen C. and W. P. Bemis. 1983. Heritability of fatty acid composition of buffalo gourd seed oil. Journal of Heredity. 74: 199–200.

Gifford, John C. 1925. Billy Bowlegs. Privately printed, Coconut Grove, Florida.

Gilmore, Melvin R. 1919. Uses of plants by the Indians of the Missouri River region. Bureau of American Ethnology Annual Reports 33: 43–154.

Gould, Fred. 1986. Simulation models for predicting durability of insect resistant germ plasm: Hessian fly (Diptera: Cecilomyiidae)—resistant winter wheat. Environmental Entomology 15: 11–23.

Great Lakes Indian Fish and Wildlife Commission. 1987. Manomin: Lake Superior gourmet wild rice. GLIFWC pamphlet, Odanah, Wisconsin.

Grossman, Bob. 1985. Overview of in-situ (unmanaged areas) and ex-situ technologies. Position paper, unpublished; Washington, D. C.

Hanna, Alfred Jackson and Kathryn Abbey Hanna. 1948. Lake Okeechobee: Wellspring of the Everglades. Bobbs-Merrill, Indianapolis.

Harlan, Jack R. 1975. Our vanishing genetic resources. Science 188: 618–621.

Harlan, Jack R. 1983. Negative trends in crop evolution. In: Conservation and Utilization of Exotic Germplasm to Improve Varieties. Report of the 1983 Plant

Breeding Research Forum. Pioneer Hi-Bred International, Des Moines, Iowa.

Harper, Francis 1958. The Travels of William Bartram: Naturalist's Edition. Yale University Press, New Haven.

Haury, Emil W. 1976. The Hohokam: Desert Farmers and Craftsmen. University of Arizona Press, Tucson.

Heiser, Jr., Charles B. 1985. Some botanical considerations of the early domesticated plants north of Mexico. In: Prehistoric Food Production in North America. Edited by Richard I. Ford. University of Michigan Museum of Anthropology Anthropological Papers 75: 57–72.

Hewett, Oliver H., editor. 1967. The Wild Turkey and Its Management. The Wildlife Society, Washington, D.C.

Hirth, Kenneth G. 1984. Trade and Exchange in Early Mesoamerica. University of New Mexico Press, Albuquerque.

Houseal, Brian, Craig MacFarland, Guillermo Archibald and Aurelio Chiari. 1985. Indigenous cultures and protected areas in Central America. Cultural Survival Quarterly 9(1): February.

Hunter, A. F. and L. W. Aarssen. 1988. Plants helping plants: new evidence indicates that beneficence is important in vegetation. Bio Science 38(1): 34–40.

Hurt, R. Douglas. 1987. Indian Agriculture in America: Prehistory to the Present. University Press of Kansas, Lawrence.

Jain, S. K., A. M. Olivieri, and J. Fernandez-Martinez. 1977. Serpentine sunflower, *Helianthus exilis*, as a genetic resource. Crop Science 17: 477–479.

Janzen, Daniel H. 1986. Guanacaste National Park: Tropical Ecological and Cultural Restoration. Editorial Universidad Estatal A Distancia, San José, Costa Rica.

Janzen, Daniel H. 1988. Tropical dry forests: the most endangered major tropical ecosystem. In: BioDiversity: Edited by E. O. Wilson and Frances M. Peter. National Academy Press, Washington, D. C. pp. 130–137.

Janzen, Daniel H. and Paul S. Martin. 1982. Neotropical anachronisms: the fruits the gomphotheres ate. Science 215: 19–27.

Jarvis, C. D. 1908. American varieties of beans. Cornell University Agricultural Experiment Station Bulletin 260.

Jenks, Albert Ernest. 1900. The wild rice gatherers of the upper Great Lakes: a study in American primitive economics. In: Nineteenth Annual Report of the Bureau of American Ethnology to the Secretary of the Smithsonian Institution. pp. 1019–1106.

Kaplan, Lawrence. 1981. What is the origin of the common bean? Economic Botany 35(2): 240–254.

Kruckeberg, Arthur R. 1986. An essay: the stimulus of unusual geologies for plant speciation. Systematic Botany 11(3): 455–463.

LaDuke, Winona. 1987. Native rice, native hands: the Ikwe Marketing Collective. Cultural Survival Quarterly II (1): 63–65.

Laferriere, Joseph E., 1986. Interspecific hybridization in sunflowers: an illustration of the importance of wild genetic resources in plant breeding. Outlook on Agriculture 15(3): 104–109.

Lawton, Harry W. and Philip J. Wilke. 1979. Ancient agricultural systems in dry regions. In: Agriculture in Semi-Arid Environments. Edited by A. E. Hall, G. H. Cannell, and H. W. Lawton. Springer-Verlag Berlin, Heidelberg. pp. 1–44.

LeBlanc, Steven A. 1985. History and environment in the Mimbres Valley. Masterkey 59(1): 18–24.

Leopold, Aldo Starker. 1944. The nature of heritable wildness in turkeys. Condor 46: 133–197.

Leopold, Estella and Harry D. MacGinitie. 1972. Development and affinities of Tertiary floras in the Rocky Mountains. In: Floristics and Paleofloristics of Asia and Eastern North America. Edited by A. Graham. Elsevier, Amsterdam. pp. 147–200.

Levi, M. Carolissa. 1956. Chippewa Indians of Yesterday and Today. Pageant Press, New York.

Lewandowski, Stephen. 1988. Dio he'ho, the three sisters on Seneca life: implications for a Native agriculture in the Finger Lakes. Agriculture and Human Values.

Lewis, N. E. and B. Lewis, editors. 1981. Lipoproteins, Atherosclerosis, and Coronary Heart Disease. Elsevier/North-Holland Biomedical Press, Amsterdam.

Libby, W. F. 1954. Chicago radiocarbon dates. IV. Science 119: 135–140.

Liebman, Matt. 1987. Polyculture cropping systems. In: Agroecology: The Scientific Basis of Alternative Agriculture. Edited by Miguel A. Altieri. Westview Press, Boulder. pp. 115–126.

Ligon, J. S. 1946. History and management of Merriam's wild turkey. University of New Mexico Publications in Biology 1:1–84.

Lumholtz, Carl. 1902. Unknown Mexico. Scribners, New York.

MacGinitie, Harry D. 1953. Fossil plants of the Florissant Beds, Colorado. Carnegie Institute of Washington Publications 599: 1–198.

Macgregor, Gordon. 1949. Attitudes of the Fort Berthold Indians regarding removal from the Garrison Reservoir site and future administration of their reservation. North Dakota History 16: 36–39.

Martin, Calvin. 1978. Keepers of the Game. University of California Press, Berkeley.

Martin, Paul S. 1986. Refuting late Pleistocene extinction models. In: Dynamics of Extinction. Edited by D. K. Elliot. John Wiley, New York. pp. 107–130.

Martin, Paul S. and Richard G. Klein. 1984. Quaternary Extinctions: A Prehistoric Revolution. University of Arizona Press, Tucson.

Maslow, Jonathan Evan. 1987. A dream of trees. Philadelphia (magazine), Philadelphia. pp. 198–285.

Masse, W. Bruce. 1981. Prehistoric irrigation systems in the Salt River Valley, Arizona. Science 214: 408–415.

May, Robert M. 1975. Stability in ecosystems: some comments. In: Unifying Concepts in Ecology. Edited by W. H. van Dobben and R. H. Lowe-McConnell. Dr. W. Junk Publishers, The Hague. pp. 161–168.

May, Robert M. and Anna Marie Lyles. 1986. Living Latin binomials: conservation biology. Nature 326: 642–643.

McDermott, John Francis. 1951. Up the Missouri with Audubon: The Journal of Edward Harris. University of Oklahoma Press, Norman.

McKusick, Charmion R. 1986. Southwest Indian Turkeys: Prehistory and Comparative Osteology. Southwest Bird Laboratory, Globe, Arizona.

Meeker, James E. and Lisa T. Ragland. 1987. The Vegetation of the Kakagon Slough and Vicinity. University of Wisconsin Natural History Museums Council, Madison.

Meyer, Roy W. 1968. Fort Berthold and the Garrison Dam. North Dakota History 35 (3 and 4): 20.

Meyer, Roy W. 1977. The Village Indians of the Upper Missouri: The Mandans, Hidatsas, and Arikaras. University of Nebraska Press, Lincoln.

Milanich, Jerald T. and Charles H. Fairbanks. 1980. Florida Archaeology. Academic Press, New York.

Miller, Judith. 1973. Genetic erosion: crop plants threatened by government neglect. Science 182: 1231–1233.

Minckley, W. L. 1973. Fishes of Arizona. Arizona Game and Fish Department, Phoenix.

Morton, Julia F. 1975. The sturdy Seminole pumpkin provides much food with little effort. Proceedings of the Florida State Horticultural Society 88: 137–142.

Myers, Norman. 1984. The Primary Source: Tropical Forests and Our Future. W. W. Norton, New York.

Nabhan, Gary P. 1978. Viable seeds from prehistoric caches? Archaeological remains in Southwestern folklore. The Kiva 43(2): 143–159.

Nabhan, Gary P. 1979. Cultivation and culture. The Ecologist. 9(8–9): 259–263.

Nabhan, Gary P. 1983. Kokopelli: the humpbacked flute player. CoEvolution Quarterly 37: 4–11.

Nabhan, Gary P. 1983. Papago fields: arid lands ethnobotany and agricultural ecology. Unpublished doctoral dissertation, University of Arizona, Tucson.

Nabhan, Gary P. 1983. Wild species protected by Arizona Hopi farmers. The Sunflower (December) 34–35.

Nabhan, Gary P. 1985. Native American crop diversity, genetic resource conservation, and the policy of neglect. Agriculture and Human Values 11(3): 14–17.

Nabhan, Gary P. 1986. How are tropical deforestation and desertification affecting plant genetic resources? Annals of Earth Stewards 5(1): 21–24.

Nabhan, Gary P. 1987. Ethnobiology and conservation: valuing diversity. Environment Southwest 519: 28–31.

Nabhan, Gary P. 1987. Saving native plants. American Land Forum (May-June): 12–14.

Nabhan, Gary P. 1987. The origins of Neotropical horticulture following megafaunal extinctions: Did humans disperse and select anachronistic fruits? Society of Ethnobiology Tenth Annual Conference Abstracts. Florida State Museum, Gainesville. Forthcoming from Journal of Ethnobiology.

Nabhan, Gary P., Cynthia Anson, Mahina Drees, and Danny Lopez. 1981. Kaicka: Seed-Saving the Papago-Pima Way. Meals for Millions/Freedom from Hunger Foundation/Southwest Program, Tucson, 34 pp.

Nabhan, Gary P. and Richard S. Felger. 1985. Wild relatives of crops: their direct uses as food. In: Plants for Arid Lands. Edited by G. E. Wickens, J. R. Goodin, and D. V. Field. George Allen and Unwin, London, pp. 19–34.

Nabhan, Gary P., A. M. Rea, K. L. Reichhardt, E. Mellink and C. F. Hutchinson. 1983. Papago influences on habitat and biotic diversity: Quitovac oasis ethnoecology. Journal of Ethnobiology 2(2): 124–143.

Nabhan, Gary P. and K. L. Reichhardt. 1983. Hopi protection of *Helianthus anomalus* Blake (Compositae), a rare sunflower. Southwestern Naturalist 28: 231–234.

Nabhan, Gary P., C. W. Weber and J. W. Berry. 1985. Variation in the composition of Hopi Indian beans. Ecology of Food and Nutrition 16: 135–152.

National Research Council. 1972. Genetic Vulnerability of Major Crops. National Academy of Sciences, Washington, D. C.

Nations, James D. 1984. The Lacadones, Gertrude Blom and the Selva Lacandona.

In: Gertrude Blom Bearing Witness. Edited by Alex Harris and Margaret Sartor. University of North Carolina Press, Chapel Hill. pp. 27–41.

Nations, James D. and Daniel I. Komer. 1984. Conservation in Guatemala: Final report presented to World Wildlife Fund, U. S. Center for Human Ecology, Austin, Texas.

Nelson, Ronald N. and Reynold P. Dahl. 1986. The wild rice industry: economic analysis of rapid growth and implications for Minnesota. In: University of Minnesota Department of Agricultural and Applied Economics Staff Paper: P86–25, St. Paul. pp. 1–68.

Oelke, Ervin A. 1974. Wild rice domestication as a model. In: Seed-bearing Halophytes as Food Plants. Edited by G. Fred Somers. University of Delaware, Newark. pp. 47–56.

Ohga, Ichiro. 1923. On the longevity of seeds of *Nelumbo nucifera*. Botanical Magazine of Tokyo 37: 88–95.

Ohga, Ichiro. 1926. On the structure of some ancient, but still viable fruits of Indian lotus, with special reference to their prolonged dormancy. Japanese Journal of Botany 3: 1–20.

Oldfield, Margery L. and Janis B. Alcorn. 1987. Conservation of traditional agroecosystems. BioScience 37(3): 199–208.

Olivieri, A. M. and S. K. Jain. 1977. Variation in *Helianthus exilis-bolanderi* complex: a reexamination. Madrono 24: 177–189.

Pellmyr, Olle and Leonard. Thuien. 1986. Insect reproduction and floral fragrances: keys to the evolution of angiosperms? Taxon 35(1): 76–85.

Pennington, Campbell W. 1969. The Tepehuan of Chihuahua: Their Material Culture. University of Utah Press, Salt Lake City.

Perron, Jacques Davy du. 1691. Perroniana et Thuana, 1667. Privately printed, Cologne, France.

Plucknett, D. L., N. J. H. Smith, J. T. Williams, and N. Murthi Anishetty. 1983. Germplasm conservation and developing countries. Science 220: 163-169.

Posey, Daniel A. 1984. A preliminary report on diversified management of tropical forest by the Kayapo Indians of the Brazilian Amazon. In: Ethnobotany in the Neotropics. Edited by G. T. Prance and J. A. Kallunki. New York Botanical Garden, Bronx. pp. 112–126.

Prescott-Allen, Christine and Robert Prescott-Allen. 1987. The First Resource: Wild Species in the North American Economy. Yale University Press, New Haven.

Priestley, David A. and Maarten A. Posthumus. 1982. Extreme longevity of lotus seeds from Pulantien. Nature 299: 148–149.

Providenti, R. et al. 1978. Resistance in feral species to six viruses infecting *Cucurbita*. Plant Disease Reporter 62(4): 326–327.

Ramírez-Acosta, José M. 1986. Importancia de colectas regionales del Jardín Agrícola Tropical del CRUSE. In: Memorias de la Segunda Reunión de Jardines Botánicos, 16–19 September, 1986, Saltillo, Coahuila, Mexico.

Ramsbottom, J. 1942. Recent work on germination. Nature 149: 658.

Rea, Amadeo M. 1980. Late Pleistocene and Holocene turkeys in the Southwest. Contributions to the Sciences from the Natural History Museum of Los Angeles County 330: 209–224.

Rea, Amadeo M. 1983. Once a River: Bird Life and Habitat Changes on the Middle Gila. University of Arizona Press, Tucson.

Regal, Philip J. 1977. Ecology and evolution of flowering plant dominance. Science 196: 622–629.

Ritzenthaler, Robert E. 1978. Southwestern Chippewa. In: Handbook of the North American Indians. Edited by Stuart Sturtevant. Smithsonian Institution, Washington, D.C. Volume 15.

Robinson, R. W. and J. T. Puchalski. 1980. Synonomy of *Cucurbita martinezii* and *C. okeechobeensis*. Cucurbit Genetics Cooperative Newsletter 3: 26–28.

Rogers, Charlie E., Tommy E. Thompson, and Gerald J. Seiler. 1982. Sunflower Species of the United States. National Sunflower Association, Bismarck, North Dakota.

Roos, Eric E. 1986. Precepts of successful seed storage. In: Physiology of Seed Deterioration. Crop Science Society of America Special Publication No. 11.

Saenger, Walter. 1982. Florissant Fossil Beds National Monument: Windows to the Past. Rocky Mountain Nature Association, Inc., Estes Park, Colorado.

San Marcos Recovery Team. 1984. San Marcos Recovery Plan. U. S. Fish and Wildlife Service, Albuquerque.

Sauer, Jonathan. 1977. The grain amaranths and their relatives: a revised taxonomic and geographic survey. Amaranth Round-up. Rodale Press, Emmaus, Pennsylvania. pp. 13–24.

Schaafsma, Polly. 1980. Indian Rock Art of the Southwest. University of New Mexico Press, Albuquerque.

Schorger, A. W. 1966. The Wild Turkey: Its History and Domestication. University of Oklahoma Press, Norman.

Shroeder, Albert H. 1968. Birds and feathers in documents relating to Indians of the Southwest. In: Collected Papers in Honor of Lyndon Lane Hargrave. Papers of the Archaeological Society of New Mexico 1(1): 95–114.

Simberloff, Daniel. 1987. Are we on the verge of a mass extinction in tropical rain forests? In: Dynamics of Extinction. Edited by David K. Elliott. Wiley-Interscience, New York. pp. 165–180.

Smail, J. Richard. 1987. Native American Agriculture: A Critical Resource. Scottsdale Community College American Indian Programs, Scottsdale, Arizona.

Small, John K. 1930. The Okeechobee gourd. Journal of the New York Botanical Garden 31: 10–14.

Smith, G. Hubert. 1972. Like-A-Fishhook Village and Fort Berthold, Garrison Reservoir, North Dakota. National Park Service Anthropological Papers 2, Washington, D. C.

Smith, Thomas Buckingham. 1848. Report of Buckingham Smith, Esquire, on His Reconnaissance of the Everglades, 1848. Senate Documents, Reports of the Committees, No. 242, Thirtieth Congress, First Session, Washington, D.C.

Smith, Thomas Buckingham, editor. 1871. Relation of Alvar Núñez Cabeza de Vaca. University Microfilms, Ann Arbor (reprint, 1966).

Spicer, Edward. 1962. Cycles of Conquest. University of Arizona, Tucson.

Spicer, Edward H. 1971. Persistent cultural systems. Science 174: 795–800.

Spicer, Edward. 1980. The Yaquis: A Cultural History. University of Arizona, Tucson.

Sweet, Muriel. 1962. Common Edible and Useful Plants of the West. Naturegraph Company, Healdsburg, CA.

Terrell, Edward E., and W. J. Wiser. 1975. Protein and lysine contents in grains of three species of wild-rice (Zizania: Gramineae). Botanical Gazette 136(3): 312–316.

Terrell, E. E., W. H. P. Emery and H. E. Beaty. 1978. Observations on Zizania texana (Texas wild-rice), an endangered species. Bulletin of the Torrey Botanical Club 105: 50–57.

Thannum, Jim. 1987. Great Lakes Indian Fish and Wildlife Commission FY 1988: Wild Rice Analysis and Enhancement Program. GLIFWC, Odanah, Wisconsin.

Tiffney, Bruce. 1984. Seed size, dispersal syndromes, and the rise of the angiosperms: evidence and hypothesis. Annals of the Missouri Botanical Garden 71: 551–576.

Toledo, Victor Manuel. 1987. La ethnobotánica en Latinoamérica: vicisitudes, contextos, desafíos. Memorias del IV Congreso Latinoamericano de Botánica. ICFES, Bogotá, Colombia: 13–34.

Trenbath, B. R. 1975. Diversify or be damned? The Ecologist 5: 76–83.

Unger, Douglas. 1984. Leaving the Land. Harper and Row, New York.

U. S. Congress, Office of Technology Assessment. 1986. Grassroots Conservation of

Biological Diversity in the United States—Background Paper #1. U. S. Government Printing Office OTA-BP-F-38, Washington, D. C.

Van Willigen, John. 1981. Applied anthropology and cultural persistence. In: Persistent Peoples: Cultural Enclaves in Perspective. Edited by George Pierre Castile and Gilbert Kushner. University of Arizona Press, Tucson. pp. 153–171.

Vennum, Jr. Thomas. 1988. Wild rice and the Ojibway people. Minnesota Historical Society, St. Paul.

Vovides, A. P. 1981. Relación de plantas Mexicanas raras o en peligro de extinción. Biótica 6: 222–228.

Wester, Horace V. 1973. Further evidence on age of ancient viable lotus seeds from Pulantien deposit, Manchuria. HortScience 8(5): 371–377.

Whealy, Kent. 1985. Garden Seed Inventory. Seed Savers Exchange, Decorah, Iowa.

Whealy, Kent, editor. 1986. The First Ten Years. Seed Savers Exchange, Decorah, Iowa.

Whittington, Stephen L. and Bennett Dyke. 1984. Simulating overkill: experiments with the Mosimann and Martin model. In: Quaternary Extinctions: A Prehistoric Revolution. Edited by Paul S. Martin and Richard C. Klein. University of Arizona Press, Tucson. pp. 451–465.

Wilkes, H. G. 1970. Teosinte introgression in the maize of Nabogame Valley. Harvard University Botanical Museum Leaflets 22: 297–310.

Wilkins, Robert P. and Wynona Huchette Wilkins. 1977. North Dakota: A Bicentennial History. W. W. Norton, New York.

Will, George F. 1947. Indian harvesting. North Dakota and South Dakota Horticulture 20(9): 131.

Will, George F. and George H. Hyde. 1917. Corn among the Indians of the Upper Missouri. William Harvey Miner Co., St. Louis.

Will, Lawrence. 1977. A Cracker History of Okeechobee. The Glades Historical Society, Belle Glade, Florida.

Williams, Lovette E., Jr. 1981. The Book of the Wild Turkey. Winchester Press, Tulsa.

Wilson, E. O. and Frances M. Peter, editors. 1988. BioDiversity. National Academy Press, Washington, D. C.

Wilson, Gilbert L. 1987. Buffalo Bird Woman's Garden: Agriculture of the Hidatsa Indians. Minnesota Historical Society Press, St. Paul.

Wilson, Gilbert Livingstone. 1917. Agriculture of the Hidatsa Indians: an Indian interpretation. University of Minnesota Studies in the Social Sciences 9: 1–129.

Winter, J. C. 1976. The Hovenweep Archaeological Project: a study of Anasazi agriculture. In: Proceedings of the First National Park Service Research Conference. National Park Service, New Orleans. pp. 127–154.

Wiseman, Fred M. 1980. The edge of the tropics: the transition from tropical to subtropical ecosystems in Sonora, Mexico. GeoScience and Man 21:141–156.

Wolfe, M. S. 1985. The current status and prospects of multiline cultivars and variety mixtures for disease resistance. Annual Review of Phytopathology 23: 251–273.

Wotowiec, Peter and Hector A. Martinez H. 1984. Estudios silviculturales con especies para producción de leña en la zona semiárida de Guatemala. CATIE/INF, Guatemala.

Yarnell, Richard A. 1977. Native plant husbandry north of Mexico. In: Origins of Agriculture. Edited by Charles A. Reed. Mouton, The Hague. pp. 861-874.

Yarnell, Richard A. In press. Sunflower, sumpweed, amaranth, devil's claw, and semi-cultigens. In: Handbook of North American Indians. Edited by William C. Sturtevant. Smithsonian Institution, Washington, D. C.

INDEX

ABOUT THE AUTHOR

An ethnobiologist and Arab American, Gary Paul Nabhan has done field research on the conservation and culinary uses of desert plants for a quarter century in the United States, Mexico, Lebanon, Oman, and Egypt. He is cofounder of Native Seeds/ SEARCH and works with native farmers throughout the southwestern United States and northern Mexico to conserve traditional food plants and their associated folklore. He has been honored with a MacArthur "Genius" Award, a Lannan Literary Fellowship, and a Lifetime Achievement Award from the Society for Conservation Biology. Nabhan is the author, editor, or coauthor of twenty books, three of which have won national or international awards. He has also served on the board of the Arab-American Writer's Guild known as al-RAWI. He was the founding director of the Center for Sustainable Environments at Northern Arizona University before returning to his desert home near Tucson, where he is now affiliated with the Southwest Center of the University of Arizona. See www.garynabhan.com.